MW00760539

DYNAMICAL MODELING IN BIOTECHNOLOGY

 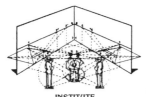

INSTITUTE
FOR SCIENTIFIC INTERCHANGE

Lectures presented at the EU Advanced Workshop on

DYNAMICAL MODELING IN BIOTECHNOLOGY

DGXII European Commission BIOTECH (1994–1998) Programme

Villa Gualino, Torino, Italy   27 May–9 June 1996

Edited by

Franco Bagnoli & Stefano Ruffo

Università di Firenze, Italy

World Scientific
Singapore • New Jersey • London • Hong Kong

Published by

World Scientific Publishing Co. Pte. Ltd.

P O Box 128, Farrer Road, Singapore 912805

USA office: Suite 1B, 1060 Main Street, River Edge, NJ 07661

UK office: 57 Shelton Street, Covent Garden, London WC2H 9HE

British Library Cataloguing-in-Publication Data
A catalogue record for this book is available from the British Library.

ISBN 981-02-3604-2

Printed in Singapore by World Scientific Printers

Foreword

This book contains the lectures given at the EU Advanced Workshop "Dynamical Modeling in Biotechnology" (Torino, May 27 – June 9 1996). The workshop has been funded by the European Commission DGXII under the program BIOTECH (1994-1998) and hosted by the Institute for Scientific Interchange (ISI).

This book appears with a considerable delay, for which we make our excuses to the authors, who have been patiently waiting for the publication of their papers. Anyway, we think that this material is still worth to publish and up to date. It covers some basic aspects of modeling, which have not been surpassed in spite of the rapid growth of this field. No effort has been made to update the articles concerning references and similar aspect, not to further delay the publication.

The investigation of chaotic dynamics, out of equilibrium statistical mechanics and disordered system has become a mature field in theoretical physics and computer science. The basic approach consists in trying to catch some essential features of a real system by building a model where the interaction of several simple elements originates a non trivial global behavior. The tools developed in this area have now a widespread application in biology.

Modelization in physics and in engineering is a growing field. However, in biotechnology, most of the theoretical studies stop at the phenomenological level. On the other side, theoretical models are often quite far from the reality of biological systems. Mathematics has long held a central position in the physical sciences, where systems, in closely controlled environments, change in a repeatable way. For biological systems, however, observations show much greater variability and thus breakthroughs and important results can only depend on a close collaborative efforts between theoretical scientists and "wet-lab" scientists. Time is now ripe for a productive interaction among biologists and theorists: the former should develop more realistic models; the latter are expected, on their side, to accept theoretical work and incorporate it into practical applications.

In this spirit we have organized this workshop, during which some applications of the theory of dynamical systems and statistical mechanics to biotechnology were presented to young researchers (master and PhD students) coming from biology, physics and mathematics.

The workshop was structured into morning lectures and afternoon computer projects followed by tutors. This allowed the immediate implementation of the models proposed during the lectures. Lecturers, tutors and students lived together two intense weeks. Lecture rooms and offices with computer facilities were made available by ISI. The first three days of the workshop were devoted to "computer alphabetization" in C and FORTRAN, covering also the use of graphics routines and of DNA databases, and to introductory lectures in theoretical biology. This period also allowed students to form several groups, which then worked at the computer projects. Ten-days projects were organized dealing with advanced topics. The home page of the workshop still exists [a] and is serving as a reference for students and lecturers to exchange materials and information. One can find there the list the lectures, projects and seminars.

A total of 30 students participated to the workshop from different European countries. Among them, 13 students were from biological or medical disciplines, the rest mostly from physics or from mathematics, engineering and computer science. The workshop has been generally appreciated. It could set a scheme for future courses on these subjects, to be held at the doctoral or post-doctoral level.

This book is divided into a first part containing the lecture notes and a second one devoted to contributed papers on specific research topics.

In the first paper, Franco Bagnoli provides a comprehensive overview of cellular automata and discusses why this approach is a flexible and powerful modelization tool for biological systems. Dietrich Stauffer presents a review of several population models. He aims at studying ageing and he shows that even simple stochastic models can be very successful in reproducing some aspects of experimental data. Giovanna Guasti explores the possibility of modeling translation process and ribosome dynamics in bacteria. This approach sheds some light on the dependence of a macroscopic quantity - bacterial population growth - on a microscopic parameter - ribosomes and tRNA abundances and composition.

Nino Boccara discusses the modeling of ecological and epidemiological systems and presents a microscopic model whose mean-field limit is represented by the "classic" Lotka-Volterra equations. Jean-Pierre Nadal presents thereafter an introduction and selected bibliography to a increasingly applied tool: neural networks and supervised learning.

The paper by Arkady Pikovsky describes the fundamentals of dynamical system theory. It offers a wide range of examples, introducing the concepts of trajectories in a phase space, stability, bifurcations, fluctuations, attractors and chaos. Numerical methods are also discussed as well as the implications

---

[a]http://www.docs.unifi.it/biotech96

of "discrete" vs. "continuum" approaches to biological modeling.

Michel Peyrard analyzes the dynamical properties of "DNA openings", which is one of the most important steps in replication and transcription. He stresses the importance of nonlinearities and how the idea of nonlinear energy localization (breathers) can be useful to interpret DNA thermal denaturation.

A general discussion of the role of modelization in biology can be found in the contribution by Franco Celada, in which he presents a comprehensive description of the immune system and suggests perspectives to its modelization.

Michele Bezzi examines some key models of the immune system and analyzes in detail the Celada-Seiden model for humoral response. Philip Seiden describes a two dimensional cellular automaton model of the dynamics of migrating, contact-inhibited cell populations. Ulrich Behn, Holger Dambeck and Gerhard Metzner present a model of the behavior of class 1 and 2 T helper lymphocytes. These cells show reciprocal inhibitory action and differ in their pattern of cytokine production. The understanding of the regulation of these two classes of T lymphocytes is thought to be of key importance in the future treatment of viral and bacterial infections and of auto-immune diseases.

In the articles section, Franco Bagnoli, Nino Boccara and Paolo Palmerini present the analysis of the behavior of a probabilistic cellular automata with two absorbing states and address questions concerning phase transitions and the analogy with percolation models. Robin Engelhardt's paper deals with pattern formation, which is a central issue in biology due to the close relation between morphology (structure) and function. The author discusses the "chemical" mechanisms of pattern formation and the biological constraints related to the presence of morphogenetic gradients during the development and reshaping processes and the control of gene expression as a source of positional information.

Peter Goetz discusses the kinetics of microbial growth in various conditions, linking the biomass production to the bacterial metabolism and growth. His model takes into account the delay time between the change of conditions in the bioreactor and the intracellular concentration of metabolites. Markus Rarbach and Peter Goetz consider the problem of mixed bacterial cultures that is nowadays of great importance in bioreactor technology in order to set up complex biotechnological processes. The authors study the stability of a system of two microbial species with commensalistic and mutualistic interactions and compare it with mathematical models of a pure culture.

Models like the Celada-Seiden one are computationally very expensive, even using powerful workstation. Filippo Castiglione, Mauro Bernaschi and Sauro Succi describe a parallel platform optimization of this model.

In summary, this collection of lectures and papers addresses biological systems modelization from a dynamical point of view, with a particular attention to biotechnology related problems. It provides an insight into a new fascinating and fast growing area of science.

We wish to thank dr. Pietro Liò, who helped us in the organization of the workshop and participated to several activities; regrettably, he decided not to join us in editing this book. We thank also dr. Alessio Vassarotti from DGXII Commission for believing in this project from the beginning and for helping us in the organization and funding process. We finally thank Prof. Mario Rasetti, director of ISI, who agreed on hosting this workshop at his institute and pushed forward the publication of this book. We acknowledge the precious help of the staff at ISI, dr. Tiziana Bertoletti and Mrs. Enza Palazzo.

Franco Bagnoli and Stefano Ruffo, Florence, August 2000

CONTENTS

Lectures

CELLULAR AUTOMATA

FRANCO BAGNOLI
Dipartimento di Matematica Applicata
Università di Firenze, via S. Marta, 3
I-50139 Firenze Italy
e-mail: bagnoli@dma.unifi.it

1 Introduction

In this lecture I shall present a class of mathematical tools for modeling phenomena that can be described in terms of elementary interacting objects. The goal is to make the macroscopic behavior arise from individual dynamics. I shall denote these individuals with the term automaton, in order to emphasize the main ingredients of the schematization: parallelism and locality. In my opinion, a *good* microscopic model is based on a rule that can be executed in parallel by several automata, each of which has information only on what happens in its vicinity (that can extend arbitrarily). In the following I shall use the word automaton either to refer to a single machine or to the whole set of machines sharing the same evolution rule.

These are completely discrete models: the time increases by finite steps, the space is represented by a regular lattice, and also the possible states of the automaton (the space variables) can assume one out of a finite set of values. The reason for this choice is the conceptual simplicity of the description. A single real number requires an infinity of information to be completely specified, and since we do not have any automatic tool (i.e. computer) that efficiently manipulates real numbers, we have to resort to approximations. On the other hand, these discrete models can be exactly implemented on a computer, which is a rather simple object (being made by humans). Since the vast majority of present computers are serial ones (only one instruction can be executed at a time), the parallel nature of the model has to be simulated.

The goal of a simple microscopic description does not imply that one cannot use real numbers in the actual calculations, like for instance in mean field approximations.

The class of phenomena that can be described with automata models is very large. There are real particle-like objects, such as atoms or molecules (from a classical point of view), that can be used to model the behavior of a gas. But one can build up models in which the automata represents bacteria in a culture or cells in a human body, or patches of ground in a forest.

3

In reality there are two classes of automata, one in which the automata can wander in space, like molecules of a gas, and another one in which the automata are stick to the cell of a lattice. I shall call the first type *molecular automata*, and the second *cellular automata.* [a]

Probably the first type is the most intuitive one, since it resembles actual robots that sniff around and move. Each class has its own advantages, in term of simplicity of the description. A molecular automaton has information about its identity and its position. It can be used to model an animal, its state representing for instance the sex and the age. It is quite easy to write down a rule to make it respond to external stimuli. However, one runs into troubles when tries to associate a finite spatial dimension to this automaton. Let us suppose that the automaton occupies a cell on the lattice that represents the space. The evolution rule has to decide what happens if more than one automaton try to occupy the same cell. This is very hard to do with a true parallel, local dynamics, and can involve a negotiation which slows down the simulation. Clearly, a possible solution is to adopt a serial point of view: choose one of the automata and let it move. This is the approach of the computations based on molecular dynamics, where one tries to study the behavior of ensembles of objects following Newton's equations, or in Monte Carlo calculations. This serial approach is justified when time is continuous, and the discretization is just a computational tool, since for a finite number of objects the probability of having two moves at the same instant is vanishing, but indeed it is not a very elegant solution. Thus, molecular automata are good for point-like, non-excluding particles.

The other solution is to identify the automata with a point in space: on each cell of the lattice there is a processor that can communicate with the neighboring ones. If we say that a state represents the empty cell and another the presence of a bacterium, we associate to it a well defined portion of space. From this point of view, the bacterium is nothing but a property of space. The usual name for this kind of models is *cellular automata.*

People from physics will realize that cellular automata correspond to a field-like point of view, while molecular automata correspond to a particle point of view. In the last example above, only one particle can sit at a certain location (cell) at a certain time. Thus, we described a Fermion field. One can allow also an arbitrary number of particles to share the same position. This is a Boson field, in which particles loose their individuality. Following our analogy with elementary particles, we could say that molecular automata correspond to classical distinguishable particles.

The Boson field represents also the link between molecular and cellular

---

[a] See also the contribution by N. Boccara, this volume.

automata. Indeed, if we relax the need of identifying each particle, and if we allow them to share their state (i.e. their identity and their position), then the two kinds of automata coincide. This suggests also an efficient way to simulate a molecular automata. Let us assume that every particle follows a probabilistic rule, i.e. it can choose among several possibilities with corresponding probabilities. Consider the case in which there are several identical particles at a certain position. Instead of computing the fate of each particle, we can calculate the number of identical particles that will follow a certain choice. If the number of identical particles at a certain position is large, this approach will speed up very much the simulation.

In the following I shall concentrate on cellular automata. They have been introduced in the forties by John von Neumann,[1] a mathematician that was also working on the very first computers. Indeed, cellular automata represent a paradigm for parallel computation, but von Neumann was rather studying the logical basis of life. The idea of the genetic code was just arising at that time, so we have to take into consideration the cultural period.

From a mathematical point of view, he realized that the reproduction process implies that the organism has to include a description of itself, that we now know to be the genetic code. On the other hand, the chemistry of real world is too complex, and also the mechanics of a robot is not simply formalized. The solution was to drastically simplify the world, reducing it to a two-dimensional lattice. The result of these studies was a cellular automaton with the power of a general-purpose computer (a Turing machine), and able to read the information to reproduce itself. This solution was quite complex (each cell could assume one out of 26 states), but now we have several simplified versions. One of these is notably based on the Game of Life, a cellular automaton described in Section 4.1. For a review of these topics, see Sigmund (1993).[2]

In spite of their mathematical interests, cellular automata has been quiescent for nearly 30 years, until the introduction of the John Conway's Game of Life in the columns of Scientific American,[3] around 1970. The Game of Life is a two-dimensional cellular automaton in which the cells can assume only two states: 0 for dead and 1 for live. Looking at the evolution of this cellular automaton an the display of a fast computer is quite astonishing. There are fermenting zones of space, and Life propagates itself by animal-like structures. From a prosaic point of view, there are interesting mathematical questions to be answered, as suggested in Section 4.1.

Finally, cellular automata exited the world of mathematicians around 1984, when the journal Physica dedicated a whole issue[4] to this topic. A good review of the first application of cellular automata can also be found in Wolfram's collection of articles.[5]

6

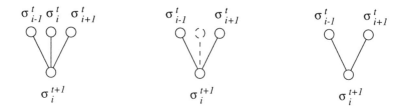

Figure 1: The $k = 1$ and $k = 1/2$ neighborhoods in $d = 1$.

In the following section I shall consider the mathematical framework of cellular automata, after which I shall review some applications of the concept.

2 Cellular Automata

Before going on we need some definitions. I shall denote the spatial index i and the temporal index t. Although we can have automata in a space of arbitrary dimension d, it is much simpler for the notations to consider a one-dimensional space. The state of a cell σ_i^t at position i and at time t can assume a finite number of states. Again for simplicity I consider only Boolean automata: $\sigma_i^t \in \{0,1\}$. Assume N as the spatial dimension of the lattice. The state of the lattice can be read as a base-two number with N digits. Let me denote it with $\sigma^t = (\sigma_1^t, \ldots, \sigma_N^t)$. Clearly $\sigma \in \{0, 2^N - 1\}$. [b]

The evolution rule $\sigma^{t+1} = F(\sigma^t)$ can in general be written in terms of a local rule

$$\sigma_i^{t+1} = f(\sigma_{i-k}^t, \ldots, \sigma_i^t, \ldots, \sigma_{i+k}^t), \tag{1}$$

where k is the range of the interactions and the boundary conditions are usually periodic. The rule f is applied in parallel to all cells. The state of a cell at time $t + 1$ depends on the state of $2k + 1$ cells at time t which constitute its neighborhood. We consider here the simplest neighborhoods: the $k = 1$ neighborhood that in $d = 1$ consists of three cells and the $k = 1/2$ neighborhood, that in $d = 1$ consists of two cells and can be considered equivalent to the previous neighborhood without the central cell. A schematic diagram of these neighborhoods is reported in Fig. 1.

---

[b]In this contribution I shall try to keep notation constant. I shall denote vectors, matrices and vectorial operators by bold symbols, including some Boolean functions of an argument that can take only a finite number of values (like a Boolean string), when referred as a whole.

Table 1: Example of Wolfram's code for rule 22.

σ_{i-1}^t	σ_i^t	σ_{i+1}^t	n	w_n	x_i^{t+1}
0	0	0	0	1	0
0	0	1	1	2	1
0	1	0	2	4	1
0	1	1	3	8	0
1	0	0	4	16	1
1	0	1	5	32	0
1	1	0	6	64	0
1	1	1	7	128	0
				total	22

2.1 Deterministic Automata

The Game of Life and von Neumann's automaton are deterministic ones, i.e. once given the initial state the fate is in principle known, even though it can take a lot of effort.

A compact way of specifying the evolution rule for $k = 1$, $d = 1$ cellular automata has been introduced by S. Wolfram.[5] It consists in reading all possible configuration of the neighborhood $(\sigma_{i-1}^t, \sigma_i^t, \sigma_{i+1}^y)$ as a base-two number n, and summing up $w_n = 2^n$ multiplied by σ_i^{t+1}, as shown in Table 1 for the rule 22.

This notation corresponds to the specification of the look-up table for the Boolean function that constitutes the evolution rule. For an efficient implementation of a cellular automata one should exploit the fact that all the bits in a computer word are evaluated in parallel. This allows a certain degree of parallelism also on a serial computer. This approach is sometimes called *multi-spin coding* (see Section 3). In the following I shall use the symbols $\oplus$, $\wedge$ and $\vee$ for the common Boolean operations eXclusive OR (XOR), AND and OR. The negation of a Boolean variable will be indicated by a line over the variable. The AND operation will be denoted often as a multiplication (which has the same effect for Boolean variables).

Let me introduce some terms that will be used in the following. If a Boolean function f of n variables $a_1, \ldots, a_n$ is completely symmetric with respect to a permutation of the variables, than it depends only on the value of the sum $\sum_i a_i$ of these variables, and it is called *totalistic*. If the function is symmetric with respect to a permutation of the variables that correspond to the values of the cells in the neighborhood, but not to the previous value of the cell, than the function depends separately on the sum of the *outer* variables

8

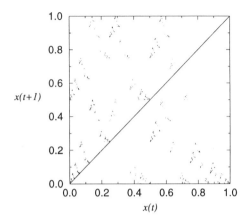

$x(t+1)$

$x(t)$

Figure 2: The return map of rule 22.

and on the previous value of the cells itself and the automaton is called *outer totalistic*.

Deterministic cellular automata are discrete dynamical systems. Given a certain state σ^t at a certain time, the state at time $t+1$ is perfectly determined. The ordered collection of states $(\sigma^0, \ldots, \sigma^t, \ldots)$ represents the trajectory of the system. Since for finite lattices the number of possible states is also finite, only limit cycles are possible. There can be several possibilities: only one cycle, several small cycles, one big and several small, etc, where big and small refer to the length of the cycle. Moreover, one can consider the basin of attraction of a cycle, i.e. the number of states that will eventually end on it.

All these dynamical quantities will change with the size of the lattice. A physicist or a mathematician is generally interested to the asymptotic behavior of the model, but there may be a well defined size more interesting than the infinite-size limit. A table with the cycle properties of the simplest cellular automata can be found at the end of Wolfram's collection.[5]

The study of the limit cycles is clearly limited to very small lattices, especially in dimension greater than one. To extend the investigation to larger lattices, one has to resort to statistical tools, like for instance the entropy of the substrings (patches) of a certain size.

If a Boolean string a, of length n appears with probability $p(a)$ (there are

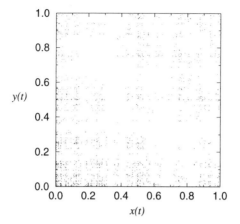

Figure 3: The attractor of rule 22.

2^n such strings), then the normalized n-entropy S_n is defined as

$$S_n = -\frac{1}{n\ln(2)} \sum_a p(a)\ln(p(a)). \tag{2}$$

S_n ranges from 1, if all strings appear with the same probability $p = 1/2^n$, to 0 if only one string appears.

One is generally interested in the scaling of S_n with n. For an example of the application of this method, see Grassberger's papers on rule 22.[6]

If one reads the configuration $\sigma^t = 01001001110101\ldots$ as the decimal digit of the fractional part of a base-two number $x(t)$ (dyadic representation), i.e.

$$x(t) = 0.01001001110101\ldots \tag{3}$$

for an infinite lattice one has a correspondence between the points in the unit interval and the configurations of the automaton. One has thus a complete correspondence between automata and maps $x(t+1) = f(x(t))$ of the unit interval. The map f is generally very structured (see Fig. 2). This correspondence helps to introduce the tools used in the study of dynamical systems.

Another possibility is the plot of the attractor and the measure of its fractal dimension. This can be performed dividing the configuration in two parts, and reading the left (right) part as a decimal number $x(t)$ $(y(t))$. The portrait of the attractor for rule 22 is given in Fig. 3.

10

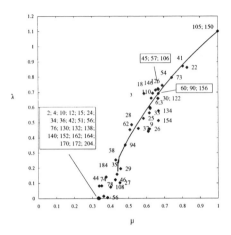

Figure 4: Lyapunov exponent for $k = 1$ cellular automata. The symbol μ denotes the instantaneous diverging rate of trajectories, and the solid line the results of a random matrix approximation. For further details, see Bagnoli *et al.* (1992).[7]

One can try to extend to these dynamical systems the concept of chaotic trajectories. In the language of dynamical systems a good indicator of chaoticity is the positivity of the maximal Lyapunov exponent, which measures the dependence of the trajectory with respect to a small change in the initial position. This concept can be extended to cellular automata in two ways. The first solution is to exploit the correspondence between configurations and point in the unit interval. At the very end, this reduces to the study of the dynamical properties of the map f. This solution is not very elegant, since it looses the democratic point of view of the system (each cell has the same importance).

Another possibility is to look at the state of the system as a point in very high dimensional Boolean space. Here the smallest perturbation is a change in just one cell, and this damage can propagate at most linearly for locally interacting automata. However, one can measure the instantaneous spreading of the damage (i.e. the spreading in one time step), and from here it is possible to calculate a Lyapunov exponent.[7] The automata that exhibit *disordered* patterns have indeed a positive exponent for almost all starting points (trajectories), while *simple* automata can have Lyapunov exponent ranging from $-\infty$ to positive values. As for deterministic dynamical systems one interprets the trajectories with negative Lyapunov exponents as stable ones, and those with a positive exponent as unstable ones. In the simulation of an usual dynamical system, rounding effects on the state variables lead to the disappearance of

space

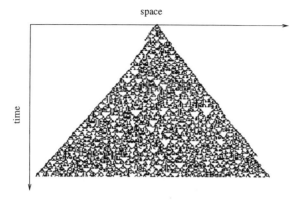

time

Figure 5: The temporal plot of the damage spreading for rule 22.

the most unstable trajectories, so that the dependence of Lyapunov exponents on the trajectories looks quite smooth. We can recover this behavior with our discrete dynamical systems by adding a small amount of noise to the evolution. A plot of the maximal Lyapunov exponent for elementary cellular automata is reported in Fig. 4.

The only source of randomness for deterministic cellular automata is in the initial configuration. The counterpart of chaoticity is the dependence on the initial condition. For chaotic automata a small change in the initial configuration propagates to all the lattice. Measures of this kind are called *damage spreading* or *replica symmetry breaking*.

It is convenient to introduce the difference field (damage) h_i^t between two configurations x_i^t and y_i^t

$$h_i^t = x_i^t \oplus y_i^t \tag{4}$$

and the Hamming distance $H(t) = 1/n \sum_{i=1}^n h_i^t$. In Fig. 5 the spreading of the difference from a single site for rule 22 is reported (see also Section 3.3).

Finally, instead of studying the diverging rate of two trajectories, one can measure the strength required to make all trajectories coalesce. Once again, this is an indicator of the chaoticity of the automaton.

2.2 Probabilistic Automata

For probabilistic cellular automata the look-up table is replaced by a table of transition probabilities that express the probability of obtaining $\sigma_i^{t+1} = 1$ once given the neighborhood configuration. For $k = 1/2$ we have four transition probabilities

$$\tau(0,0 \to 1) = p_0;$$
$$\tau(0,1 \to 1) = p_1;$$
$$\tau(1,0 \to 1) = p_2;$$
$$\tau(1,1 \to 1) = p_3,$$

(5)

with $\tau(a,b \to 0) = 1 - \tau(a,b \to 1)$.

The evolution rule becomes a probabilistic Boolean function. This can be written as a deterministic Boolean function of some random bits, thus allowing the use of multi-site coding. In order to do that, one has to write the deterministic Boolean functions that characterize the configurations in the neighborhood with the same probability. For instance, let us suppose that the $k = 1/2$ automaton of equation (5) has to be simulated with $p_0 = 0$, $p_1 = p_2 = p$ and $p_3 = q$. The Boolean function that gives 1 only for the configurations $(\sigma_{i-1}^t, \sigma_{i+1}^t) = (0,1)$ or $(1,0)$ is

$$\chi_i^t(1) = \sigma_{i-1}^t \oplus \sigma_{i+1}^t$$

(6)

and the function that characterizes the configuration $(1,1)$ is

$$\chi_i^t(2) = \sigma_{i-1}^t \wedge \sigma_{i+1}^t.$$

(7)

Let me introduce the truth function. The expression $[\![expression]\!]$ gives 1 if *expression* is true and zero otherwise (it is an extension of the delta function). Given a neighborhood configuration $\{\sigma_{i-1}^t, \sigma_{i+1}^t\}$, σ_i^{t+1} can be 1 with probability p_i. This can be done by extracting a random number $0 \le r < 1$ and computing

$$\sigma_i^{t+1} = \sum_i [\![p_i > r]\!] \chi_i.$$

(8)

Although the sum is not a bitwise operation, it can safely used here since only one out of the χ_i can be one for a given configuration of the neighborhood. The sum can be replaced with a OR or a XOR operation.

Given a certain lattice configuration $a = \sigma^t$ at time t, the local transition probabilities allow us to compute the transition probability T_{ba} from configuration b to configuration a as

$$T_{ba} = \prod_{i=1}^{N} \tau(a_{i-1}, a_{i+1} \to b_i).$$

(9)

The factorization of the matrix T implies that it can be written as a product of simpler transfer matrices M that add only one site to the configuration.

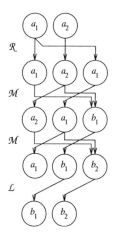

Figure 6: The application of the transfer matrix for the $k = 1/2$, $d = 1$, $N = 2$ cellular automata.

Periodic boundary conditions require some attention: in Fig. 6 an example of decomposition $T = LM^N R$ for the simplest case $N = 2$ is shown. The lattice has been skewed for the ease of visualization.

This transition matrix defines a Markov process.

The state of the system at a certain time t is indicated by the probability $x_a^{(t)}$ of observing the configuration a. Clearly, one has $\sum_a x_a^{(t)} = 1$. For a deterministic automata only one component of x is one, and all other are null. The time evolution of x is given by the applications of the transfer matrix T. The conservation of probability implies that $\sum_a T_{ba} = 1$. It is possible to prove that the maximum eigenvalue λ_1 of T is one; the eigenvector corresponding to this eigenvalue is an asymptotic state of the system.

If all configurations are connected by a chain of transition probabilities (i.e. there is always the possibility of going from a state to another) than the asymptotic state is unique.

The second eigenvalue λ_2 gives the correlation length ξ:

$$\xi = -(\ln \lambda_2)^{-1}. \tag{10}$$

In our system two different correlation length can be defined: one in the space direction and one in the time direction. The temporal correlation length gives the characteristic time of convergence to the asymptotic state.

The usual equilibrium spin models, like the Ising or the Potts model, are equivalent to a subclass of probabilistic cellular automata. In this case the

$$
M = \begin{pmatrix}
1 & 1-p & 0 & 0 & 0 & 0 & 0 & 0 \\
0 & 0 & 1-p & 1-q & 0 & 0 & 0 & 0 \\
0 & 0 & 0 & 0 & 1 & 1-p & 0 & 0 \\
0 & 0 & 0 & 0 & 0 & 0 & 1-p & 1-q \\
0 & p & 0 & 0 & 0 & 0 & 0 & 0 \\
0 & 0 & p & q & 0 & 0 & 0 & 0 \\
0 & 0 & 0 & 0 & 0 & p & 0 & 0 \\
0 & 0 & 0 & 0 & 0 & 0 & p & q
\end{pmatrix}
$$

Figure 7: Transfer Matrix M for $N = 2$.

transition probability between two configurations a and b is constrained by the detailed balance principle

$$
\frac{T_{ba}}{T_{ab}} = \exp\left(\beta H(b) - \beta H(a)\right), \tag{11}
$$

where $H(a)$ is the energy of configuration a and β is the inverse of the temperature. One can invert the relation between equilibrium spin models and probabilistic cellular automata [8] and reconstruct the Hamiltonian from the transition probabilities. In general, a cellular automata is equivalent to an equilibrium spin system if no transition probabilities are zero or one. Clearly in this case the space of configurations is connected.

These systems can undergo a phase transition, in which some observable displays a non-analytic behavior in correspondence of a well defined value of a parameter. The phase transition can also be indicated by the non-ergodicity of one phase. In other words in the phase where the ergodicity is broken the asymptotic state is no more unique. An example is the ferromagnetic transition; in the ferromagnetic phase there are two ensembles of states that are not connected in the thermodynamic limit. In the language of the transfer matrix this implies that there are two (or more) degenerate eigenvalues equal to one. The corresponding eigenvectors can be chosen such that one has positive components (i.e. probabilities) for the states corresponding to one phase, and null components for the states corresponding to the other phase, and vice versa. Indeed, the degeneration of eigenvalues correspond to the divergence of the (spatial and temporal) correlation lengths.

It is well known that equilibrium models cannot show phase transitions at a finite temperature (not zero nor infinite) in one dimension. However, this is no more true if some transition probabilities is zero or one violating the detailed balance Eq. (9). In effect, also the one dimensional Ising model exhibits a phase transition at a vanishing temperature, i.e. when some transition probabilities become deterministic. In particular one can study the phase transitions of models with adsorbing states, that are configurations corresponding to attracting points in the language of dynamical systems. An automata with adsorbing states is like a mixture of deterministic and probabilistic rules. A system with adsorbing states cannot be equivalent to an equilibrium model.

Let me introduce here an explicit model in order to be more concrete: the Domany-Kinzel model. [9] This model is defined on the lattice $k = 1/2$ and the transition probabilities are

$$\tau(0,0 \to 1) = 0;$$
$$\tau(0,1 \to 1) = p;$$
$$\tau(1,0 \to 1) = p;$$
$$\tau(1,1 \to 1) = q.$$

(12)

The configuration 0, in which all cells assume the value zero, is the adsorbing state. Looking at the single site transfer matrix M (reported in Fig. 7 for the simple case $N = 2$), one can see that (for $p, q < 1$) every configuration has a finite probability of going into the configuration 0, while one can never exit this state. In this simple probability space a state (i.e. a vector) $v = (v_0, \ldots, v_7)$ that corresponds to the single configuration a is given by $v_a = \delta_{ab}$. The configuration 0 corresponds to the state given by the vector $w^{(1)} = (1, 0, 0, \ldots)$.

Let me indicate with the symbols $\lambda_i, i = 1, \ldots, 8$ the eigenvalues of M, with $||\lambda_1|| > ||\lambda_2|| > \ldots > ||\lambda_8||$. The corresponding eigenvectors are $w^{(1)}$, $\ldots, w^{(8)}$. Let us suppose that they form a base in this space. The Markovian character of M implies that the maximum eigenvalue is 1, corresponding to the eigenvector $w^{(1)}$.

A generic vector v can be written as

$$v = a_1 w^{(1)} + a_2 w^{(2)} + \cdots$$

(13)

If we start from the vector v at time $t = 0$ and we apply T times the transfer matrix T, in the limit of large N this is practically equivalent to the application of the matrix M NT times, and we get

$$v(T, N) = M^{NT} v(0, N) = a_1 \lambda_1^{NT} w^{(1)} + a_2 \lambda_2^{NT} w^{(2)} + \cdots$$

(14)

and since all eigenvalues except the first one are in norm less that one, it follows that the asymptotic state is given by the vector $w^{(1)}$ (i.e. by the absorbing configuration 0).

This situation can change in the limit $N \to \infty$ and (after) $T \to \infty$ (the thermodynamic limit). In this limit some other eigenvalues can degenerate with

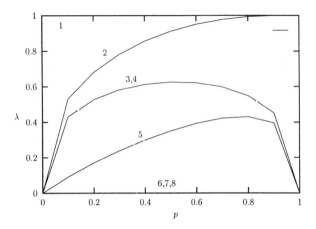

Figure 8: Modulus of the eigenvectors of the $N = 2$ transfer matrix for $p = q$.

λ_1, and thus one or more configurations can survive forever. The existence of such phase transitions can be inferred from the plot of the modulus of the eigenvalues of M for $N = 2$. This is reported in Fig. 8 for the case $p = q$. In this finite lattice the degeneration with the eigenvalue λ_1 (which corresponds to the configuration in which all sites take value 1) occurs for $p = 1$. As discussed in Section 4.2 in the thermodynamic limit the transition occurs at $p = 0.705\ldots$. A rather different scenario exhibits for $q = 0$. In this case, as reported in Fig. 9, all eigenvalues degenerate with the first (for $p = 0.81\ldots$ in the thermodynamic limit) and this implies a different dynamical behavior (see Section 4.2).

Critical Phenomena

The already cited phase transitions are a very interesting subject of study by itself. We have seen that the correlation length ξ diverges in the vicinity of a phase transition. The correlation between two sites of the lattice at distance r is supposed to behave as $\exp(-r/\xi)$, the divergence of ξ implies very large correlations. Let us suppose that there is a parameter p that can be varied. In the vicinity of the critical value p_c of this parameter the correlation length diverges as $\xi \sim (p - p_c)^{-\nu}$, where ν is in general non integer. Also other quantities behaves algebraically near a critical point, like for instance the magnetization ρ which scales as $\rho \sim (p - p_c)^{\beta}$.

This power-law behavior implies that there is no characteristic scale of

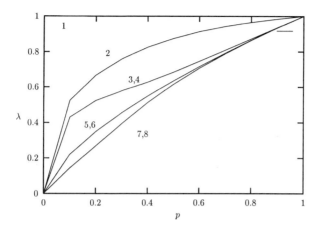

Figure 9: Modulus of the eigenvectors of the $N = 2$ transfer matrix for $q = 0$.

the phenomena. Indeed, if we change the scale in which the parameter p is measured, the proportionality factor will change but the form of the law will not. The pictorial way of explaining this phenomena is the following: suppose we have some two-dimensional system near to a phase transition where one phase is white and the other is black. There will be patches and clusters of all sizes. If we look at this specimen by means of a TV camera, the finest details will be averaged on. Now let us increase the distance from the TV camera and the specimen (of infinite extension). Larger and larger details will be averaged out. If the system has a characteristic scale ξ, the picture will change qualitatively when the area monitored will be of order of the square of this length or more. On the other hand, if we are unable to deduce the distance of the TV camera from the surface by looking at the image, the system is self-similar and this is a sign of a critical phenomena. Another sign is the slow response to a stimulation: again, the distribution of the response times follows a power-law.

The critical phenomena are very sensible to the parameter p, and a small change in it will destroy this self-similarity. In spite of their non-robustness, critical phenomena are heavily studied because their universality: since only large scale correlations are important, the details of the rule do not change the exponents of the power laws. This implies that the critical points of very different systems are very similar.

Self-similar objects are very common in nature. One example is given by

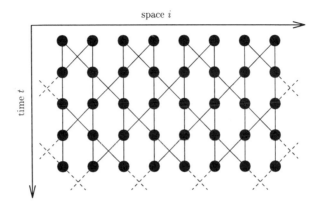

Figure 10: The lattice for the diffusion in $d = 1$.

clouds: it is almost impossible to evaluate the distance from a cloud, even when one is relatively near to it. And also power laws are found in many different fields, so are slow response times. It is impossible that we always meet usual critical phenomena, which are so delicate. It has been proposed [10] that open systems (i.e. out of equilibrium system) can auto-organize into a self-organized critical state, a state that exhibits the characteristic of a critical point still being stable against perturbations. It is indeed possible to develop simple models that exhibit this feature, but it is not clear how ubiquitous this quality is. The Game of Life of Section 4.1 is a very simple example of a Self Organized Critical (SOC) model.

Diffusion

The implementation of diffusion in the context of cellular automata is not very straightforward. One possibility is of course that of exchanging a certain number of pairs of cells randomly chosen in the lattice. While this approach clearly mixes the cells, [c] it is a serial method that cannot be applied in parallel. A more sophisticated technique is that of dividing the lattice in pieces (a one dimensional lattice can be divided in couples of two cells, a two dimensional lattice in squares of four cells), and then rotate the cell in a patch according with a certain probability. This is equivalent to considering more complex

---

[c]For an application of this technique, refer to the contribution by Boccara, this volume.

lattices, but always with a local character, as shown in Fig. 10 for $d = 1$. The diffusion can be controlled by a parameter D that gives the probability of rotating the block. One has also to establish how many times the procedure has to be repeated (changing the blocks). A detailed analysis of the effects of this procedure can be found in Chopard and Droz (1989).[11]

3 Numerical techniques

In this section I shall review some techniques that I currently use to investigate cellular automata.

3.1 Direct simulations

The first approach is of course that of assigning each cell in the lattice to a computer word, generally of type integer, and to write the evolution rule using if... instructions. This generally produces very slow programs and waste a large amount of memory (i.e. the size of the lattices are limited).

Another possibility is to use look-up tables. In this case one adds the value of the cells in the neighborhood multiplying them by an appropriate factor (or packs them into the bits of a single word), and then uses this number as an index in an array built before the simulation according with the rule.

Alternatively, one can store the configurations in patches of 32 or 64 bits (the size of a computer word). This implies a certain degree of gymnastic to make the patches to fit together at boundaries.

Again, since often one has to repeat the simulation starting from different initial configurations, one can work in parallel on 32 or 64 replicas of the same lattice, with the natural geometry of the lattice. When using probabilistic cellular automata (i.e. random bits), one can use the same input for all the bits, thus making this method ideal for the damage spreading investigations (see Section 3.3).

Finally, one can exploit the natural parallelism of serial computers to perform in parallel simulations for the whole phase diagram, as described in Bagnoli et al. (1997).[12]

In order to use the last three methods, all manipulations of the bits have to be done using the bitwise operations AND, OR, XOR. It is possible to obtain a Boolean expression of a rule starting from the look-up table (canonical form), but this generally implies many operations. There are methods to reduce the length of the Boolean expressions.[13,14]

3.2 Mean Field

The mean field approach is better described using probabilistic cellular automata. Given a lattice of size L, its state is defined by the probability x_a of observing the configuration a. If the correlation length ξ is less than L, two cell separated by a distance greater that ξ are practically independent. The system acts like a collection of subsystems each of length ξ. Since ξ is not known a priori, one assumes a certain correlation length l and thus a certain system size L, and computes the quantity of interest. By comparing the values of these quantities with increasing L generally a clear scaling law appears, allowing to extrapolate the results to the case $L \to \infty$.

The very first step is to assume $l = 1$. In this case the $l = 1$ cluster probabilities are $\pi_1(1)$ and $\pi_1(0) = 1 - \pi(1)$ for the normalization. The $l = 1$ clusters are obtained by $l = 1 + 2k$ clusters via the transition probabilities.

For the $k = 1/2$ Domany-Kinzel model we have

$$
\begin{aligned}
\pi_1'(1) &= (\pi_2(0,1) + \pi_2(1,0)p + \pi_2(1,1)q. \\
\pi_1'(0) &= \pi_2(0,0) + (\pi_2(0,1) + \pi_2(1,0))(1-p) + \pi_2(1,1)(1-q),
\end{aligned}
\tag{15}
$$

where $\pi = \pi^{(t)}$ and $\pi' = \pi^{(t+1)}$.

In order to close this hierarchy of equation, one factorizes the $l = 2$ probabilities. If we call $\rho = \pi_1(1)$, we have $\pi_2(0,1) = \pi_2(1,0) = \rho(1 - \rho)$ and $\pi_2(1,1) = \rho^2$. The resulting map for the density ρ is

$$
\rho' = 2p\rho + (q - 2p)\rho^2.
\tag{16}
$$

The fixed points of this map are $\rho = 0$ and $\rho = 2p/(2p - q)$. The stability of these points is studied by following the effect of a small perturbation. We have a change in stability (i.e. the phase transition) for $p = 1/2$ regardless of q.

The mean field approach can be considered as a bridge between cellular automata and dynamical systems since it generally reduces a spatially extended discrete system to a set of coupled maps.

There are two ways of extending the above approximation. The first is still to factorize the cluster probabilities at single site level but to consider more time steps, the second is to factorize the probabilities in larger clusters. The first approach applied for two time steps implies the factorization of $\pi_3(a,b,c) = \pi_1(a)\pi_1(b)\pi_1(c)$ and the map is obtained by applying for two time steps the transition probabilities to the $l = 3$ clusters. The map is still expressed as a polynomial of the density ρ. The advantage of this method is that we still work with a scalar (the density), but in the vicinity of a phase transition the convergence towards the thermodynamic limit is very slow.

The second approach, sometimes called *local structure approximation,* [15] is a bit more complex. Let us start from the generic l cluster probabilities π_l. We generate the $l-1$ cluster probabilities π_{l-1} from π_l by summing over one variable:

$$\pi_{l-1}(a_1, \ldots, a_{l-1}) = \sum_{a_l} \pi_l(a_1, \ldots, a_{l-1}, a_l). \tag{17}$$

The $l+1$ cluster probabilities are generated by using the following formula

$$\pi_{l+1}(a_1, a_2, \ldots, a_l, a_{l+1}) = \frac{\pi_l(a_1, \ldots, a_l)\pi_l(a_2, \ldots, a_{l+1})}{\pi_{l-1}(a_2, \ldots, a_l)}. \tag{18}$$

Finally, one is back to the l cluster probabilities by applying the transition probabilities

$$\pi'(a_1, \ldots, a_l) = \sum_{b_1, \ldots, b_{l+1}} \prod_{i=1}^{l} \tau(b_i, b_{i+1} \to a_i). \tag{19}$$

This last approach has the disadvantage that the map lives in a high-dimensional (2^l) space, but the results converges much better in the whole phase diagram.

This mean field technique can be considered an application of the transfer matrix concept to the calculation of the the eigenvector corresponding to the maximum eigenvalue (fundamental or ground state).

3.3 Damage spreading and Hamming distance

In continuous dynamical systems a very powerful indicator of chaoticity is the Lyapunov exponent. The naive definition of the (maximum) Lyapunov exponent is the diverging rate of two initially close trajectories, in the limit of vanishing initial distance.

This definition cannot be simply extended to discrete systems, but we can define some quantities that have a relation with chaoticity in dynamical systems.

First of all we need a notion of distance in discrete space. The natural definition is to count the fraction of corresponding cells that have different value (I consider here only Boolean cellular automata). If we indicate with x and y the two configurations, the distance h between them (called the Hamming distance) is

$$h = \frac{1}{n} \sum_{i=1}^{n} x_i \oplus y_i. \tag{20}$$

It is possible to define an equivalent of a Lyapunov exponent [7] (see Fig. 4), but the natural application of the Hamming distance is related to the damage spreading.

For deterministic cellular automata the damage is represented by the Hamming distance between two configurations, generally starting from a small number of damaged sites, and the goal is to classify the automata according with the average speed of the damage or other dynamical quantities.

The extension to probabilistic cellular automata is the following: what will happen if we play again a rule with a different initial configuration but *the same realization of the noise*? If the asymptotic state changes, than the evolution remembers the initial configuration, otherwise it is completely determined by the noise. In the language of dynamical systems, this two scenarios correspond to a breaking of replica (the two configurations) symmetry. We can also define the difference field $h_i = x_i \oplus y_i$ with h representing its density. The breaking of replica symmetry thus corresponds to a phase transition for h from the adsorbing state $h = 0$ to a *chaotic* state.

4 Investigation Themes

The theory described in the last section gives us the analytic tools and constitutes an interesting subject of study by itself. However, cellular automata can be studied as phenomenological models that mimic the real world from a mesoscopic point of view. The point of view of a physicist is again that of looking for the simplest model still able to reproduce the qualitative behavior of the original system. For this reason the models described in this section will be very crude; if one wants a model that mimics the system as good as possible, one can start from a simplified model and add all the features needed. An advantage of cellular automata with respect to system of differential or partial differential equations is the stability of dynamics. Adding some feature or interactions never leads to structural instabilities.

This section is of course not exhaustive.

4.1 Life

The Game of Life was introduced in the '70s by John Conway and then popularized by Martin Gardner in the columns of Scientific American. [3] It is a two dimensional, Boolean, outer totalistic, deterministic cellular automata that in some sense resembles the evolution of a bacterial population. The value 0 is associated to an empty or dead cell, while the value 1 to a live cell. The automaton is defined on a square lattice and the neighborhood is formed by the nearest and next-to-nearest neighbors. The evolution rule is symmetric in the

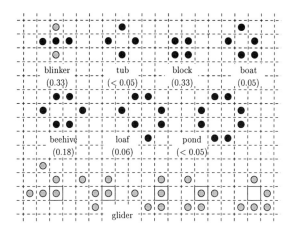

Figure 11: The most common animals in Life. The figures represent the average abundance of each animal.

values of the cells in the outer neighborhood, i.e. it depends on their sum. The transition rules are stated in a pictorial way as

- If a live cell is surrounded by less than two live cell, it will die by isolation.

- If a live cell is surrounded by more that three live cells, it will die by overcrowding.

- Otherwise, if a live cell is surrounded by two or three live cells, it will survive.

- An empty cell surrounded by three live cells will become alive.

- Otherwise an empty cell will stay empty.

The evolution of this rule is impressive if followed on the screen of a fast computer (or a dedicated machine). Starting from a random configuration with half cell alive, there is initially a rapid drop in the density ρ^t, followed by a long phase of intense activity. [d] After some hundreds time steps (for lattices of size order 200×200) there will emerge colonies of activity separated by nearly empty patches. In these patches there are small stable or oscillating configurations (the animals, see Fig. 4.1). Some of these configurations can propagate in the lattice (the gliders). The activity zones shrink very slowly,

---

[d]An exhaustive analysis of the Game of Life can be found in Bagnoli et al (1992). [16]

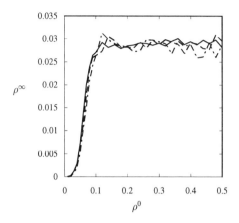

Figure 12: The plot of the asymptotic density ρ^∞ vs. the initial density ρ^0 in the Game of Life. The curves represents different lattice sizes: 96×50 (broken line), 160×100 (dashed line), 320×200 (full line).

and sometimes a glider will inoculate the activity on a quiescent zone. Finally, after hundreds time steps, the configuration will settle in a short limit cycle, with density $\rho \simeq 0.028$. The scaling of the relaxation time with the size of the lattice suggests that in an infinite lattice the activity will last forever (see Fig. 14). The existence of a non-vanishing asymptotic density has been confirmed by recent simulations performed by Gibbs and Stauffer (1997)[17] on very large lattices (up to 4×10^{10} sites).

There are several questions that have been addressed about the Game of Life. Most of them are mathematical, like the equivalence of Life to a universal computer, the existence of *gardens of heaven*, the feasibility of a self-reproducing structure in Life, and so on (see Sigmund (1993)[2]).

From a statistical point of view, the main question concerns the dependence of the asymptotic state from the density of the initial configuration (supposed uncorrelated) and the response time of the quiescent state.

For the first question, in Fig. 12 is reported the plot of the asymptotic density versus the initial density for a lattice square lattice with $L = 256$. One can see a transition of the value asymptotic density with respect to the initial density for a value of the latter around 0.05. In my opinion this effect is due to the finite size of the lattice, since there is a slight dependence on the system size. However, recent simulations performed by Stauffer[18] on very large lattices still exhibit this effect.

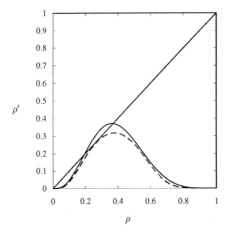

Figure 13: The mean field approximations for the Game of Life.

The response time can be measured by letting Life to relax to the asymptotic state, and then perturbing it in one cell or adding a glider. Recording the time that it takes before relaxing again to a periodic orbit and plotting its distribution, Bak et al. (1989)[19] found a power law, characteristic of self organized critical phenomena. One can investigate also the dynamical properties of the Game of Life, for instance the spreading of damage. Its distribution will probably follow a power law.

Another possibility is to investigate the asymptotic behavior using mean field techniques. The simplest approximations are unable to give even a rough approximation of the asymptotic density $\rho \simeq 0.028$, so it is worth to try more sophisticated approximations as the local structure one.

4.2   Epidemics, Forest Fires, Percolation

We can call this class of processes contact processes. They are generally representable as probabilistic cellular automata. Let me give a description in terms of an epidemic problem (non lethal)

In this models a cell of the lattice represents an individual (no empty cells), and it can stay in one of three states: healthy and susceptible (0), ill and infective (1), or immune (2).

Let us discuss the case without immunization. This simple model can be studied also in one spatial dimension (a line of individual) + time.

26

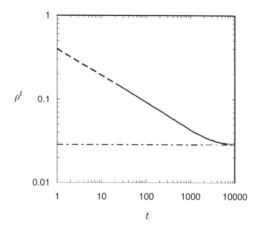

Figure 14: The temporal behavior of the density in the Game of Life.

Clearly the susceptible state is adsorbing. If there are no ill individual, the population always stays in the state 0. We have to define a range of interactions. Let us start with $k = 1$, i.e. the state of one individual will depend on its previous state and on that of its nearest neighbors. There is an obvious left-right symmetry, so that the automata is outer totalistic. The probability of having $\sigma_i^{t+1} = 1$ will depend on two variables: the previous state of the cell σ_i^t and the sum of the states of the neighboring cells. Let me use the totalistic characteristic functions $\chi_i^t(j)$ that take the value one if the sum of the variables in the neighborhood (here σ_{i-1}^t and σ_{i+1}^t) is j and zero otherwise (see Eq. (6, 7)). Moreover, I shall denote σ_i^t with σ' and I shall neglect to indicate the spatial and temporal indices. Then

$$\sigma' = f(\sigma, \chi(j)). \tag{21}$$

The function f gives the probability that σ' is ill (1) given its present state and the number of ill neighbors. If f does not depend on σ (i.e. the probability of contracting the infection is the same of staying ill for a given number of ill neighbors) we have again the Domany-Kinzel model Eq. (12). The parameters of this model are here the probability of contracting the infection or staying ill when an individual is surrounded by one (p) or two (q) sick individuals. The phase diagram of the model is reported in Fig. 15.

As one can see, there are two phases, one in which the epidemics last forever (*active* phase) and one in which the asymptotic state is only formed

27

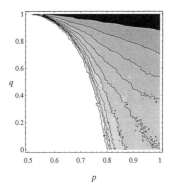 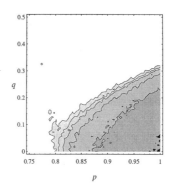

Figure 15: The phase diagram of the density (left) and damage (right) for the Domany-Kinzel model. ($0 \leq p \leq 1$, $0 \leq q \leq 1$, $N = 1000$, $T = 1000$). The gray intensity is proportional to the value of the density, ranging from white to black in equally spaced intervals.

by healthy individuals (*quiescent* phase).

A feature of this model is the presence of a region in the phase diagram with spreading of damage (*active* phase). This zone corresponds to $p \gg q$, i.e. it is due to an interference effect between the neighbors.[e] This is obviously not realistic for a generic infection, but it can have some application to the spreading of a political idea.

One can also study the case in which if an individual is surrounded by a totality of ill individuals, it will certainly contract the infection ($q = 1$). This means that also the state in which all the cells have value one is adsorbing. In this case the transition is sharp (a first order transition).

The phenomenology enriches if we allow the previous value of the cell σ to enter the evolution function. Let us consider the a simplest case: a totalistic function of the three cells. Let us call p, q and w the probability of having $\sigma' = 1$ if $\chi(1)$, $\chi(2)$ or $\chi(3)$ is one, respectively. If we set $w = 1$ we have two adsorbing states and two parameters, so that the phase diagram is again two dimensional and easy to visualize (see Fig. 16). This model will be studied in detail in the contribution by Bagnoli, Boccara and Palmerini, this volume.

We see that in this case we can have both first and second order phase transitions. This observation could have some effective importance, in that a first order phase transition exhibit hysteresis, so that if the system enters one adsorbing state, it would cost a large change in the parameter to have it switch to the other state. On the other hand, a second order phase transition is a

---

[e]See also Bagnoli (1996).[20]

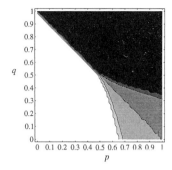 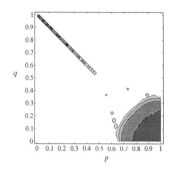

Figure 16: The phase diagram for the $k = 1$ totalistic cellular automaton with two adsorbing states.

continuous one, with no hysteresis.

The model can be extended to larger neighborhoods, immune states, more dimensions and diffusion, always looking to the phase diagram and to damage spreading. The mathematical tools that can be used are direct simulations (with the fragment technique) and various mean field approximations (once more, the local structure technique is quite powerful).

4.3 Ecosystems

I shall use the following definition of an ecosystem: an evolving set of interacting individuals exposed to Darwinian selection. We start from the simplest model: let us consider an early ecosystem, composed by haploid individuals (like bacteria). Each individual can sit on a cell of a lattice in d space dimensions. Each individuals is identified by its genetic information, represented as an integer number x. The genotype can be read as a four symbol (bases) or codon string. One could also consider the genome composed by alleles of a set of genes. I shall use here a binary coding (say, two alleles), because it is simpler to describe. All individuals have a genome of the same length l. Thus, we can interpret x as a base-two number representing the genotype or as an abstract index. The difference enters when we consider the mutations and thus we have to introduce the concept of distance in the genetic space. [f] In the first case the genotype space is an hypercube with 2^l dimensions, in the second it is an abstract space of arbitrary dimension, that for simplicity

---

[f] In effects, the distance in the genetic space is defined in terms of the number of mutations needed to connect (along the shortest path) two individuals (or, more loosely, two species).

we can consider one dimensional. Finally, we have to introduce the phenotipic distance, which is the difference in phenotipic traits between two individuals. Given a genotype x, the phenotype is represented as $g(x)$. In order to simplify the notations, I shall denote a function of the phenotype (say $h(g(x))$ as $h[x]$. The phenotypic distance will be denoted as $\delta(g(x), g(y)) = \delta[x, y]$. I do not consider the influence of age, i.e. a genotype univocally determines the phenotype. There could be more than one genotype that give origin to the same phenotype (polymorphism).

This automaton has a large number of states, one for each different genome plus a state $(*)$ for representing the empty cell. Thus $*$ represents an empty cell, 0 is the genome $(0, 0, \ldots, 0)$, 1 is the genome $(0, \ldots, 0, 1)$ and so on. The evolution of the automaton is given by the application of three rules: first of all the interactions among neighbors are considered, giving the probability of surviving, then we perform the diffusion step and finally the reproduction phase. I shall describe the one (spatial) dimensional system, but it can be easily generalized.

survival: An individual at site i in the state $x_i^t \neq *$ and surrounded by $2k + 1$ (including itself) neighbors in the states $\{x_{i-k}, \ldots, x_{i+k}\} = \{x_i\}_k$ has a probability $\pi(x_i, \{x_i\}_k)$ of surviving per unit of time. This probability is determined by two factors: a fixed term $h[x_i]$ that represents its ability of surviving in isolation, and an interaction term $1/(2k+1)\sum_{j=i-k}^{i+k} J[x_i, x_j]$. Chearly, both the static fitness and the interaction term depend on the phenotype of the individual.

The fixed field h and the interaction matrix J define the chemistry of the world and are fixed (at least in the first version of the model). The idea is that the species x with $h[x] > 0$ represent autonomous individuals that can survive in isolation (say, after an inoculation into an empty substrate), while those with $h[x] < 0$ represents predators or parasites that necessitate the presence of some other individuals to survive. The distinction between autonomous and non-autonomous species depends on the level of schematization. One could consider an ecosystem with plants and animals, so that plants are autonomous and animals are not. Or one could consider the plants as a substrate (not influenced by animals) and then herbivores are autonomous and carnivores are not. The interaction matrix specifies the necessary inputs for non autonomous species.

We can define the fitness H [9] of the species x_i in the environment $\{x_i\}_k$

---

[9]There are several definition of fitness, the one given here can be connected to the growth rate A of a genetically pure population by $A = \exp(H)$; see also Section 4.4

as

$$H(x_i, \{x_i\}_k) = h[x_i] + \frac{1}{2k+1} \sum_{j=i-k}^{i+k} J[x_i, x_j]. \tag{22}$$

The survival probability $\pi(H)$ is given by some sigma-shaped function of the fitness, such as

$$\pi(H) = \frac{e^{\beta H}}{1 + e^{\beta H}} = \frac{1}{2} + \frac{1}{2} \tanh(\beta H), \tag{23}$$

where β is a parameter that can be useful to modulate the effectiveness of selection.

The survival phase is thus expressed as:

- If $x_i \neq *$ then

$$\begin{array}{ll} x_i' = x_i & \text{with probability } \pi(H(x_i, \{x_i\}_j)) \\ x_i' = * & \text{otherwise} \end{array} \tag{24}$$

- Else

$$x_i' = x_i = * \tag{25}$$

diffusion: The diffusion is given by the application of the procedure described in Section 2.2. The disadvantage of this approach is that we cannot introduce nor an intelligent diffusion (like escaping from predators or chasing for preys), nor different diffusion rates for different species. An alternative could be the introduction of diffusion on a pair basis: two cells can exchange their identity according with a given set of rules. [h]

For the moment we can use the following rule:

- Divide the lattice in neighboring pairs (in $d = 1$ and no dependence on the neighbors there are two ways of doing that)

- Exchange the values of the cells with probability D

Clearly the probability D could depend on the value of the cells and on that of neighbors.

---

[h]There are several ways of dividing a lattice into couples of neighboring cells. If one want to update them in parallel, and if the updating rule depends on the state of neighboring cells, one cannot update a couple at the same time of a cell in the neighborhood. Since one has to update a sublattice after another, there is a slightly asymmetry that vanishes in the limit of a small diffusion probability iterated over several sublattices.

reproduction: The reproduction phase can be implemented as a rule for empty cells: they choose one of the neighbors at random and copy its identity with some errors, determined by a mutational probability ω.

Thus

- If the cell has value $*$ then
  - choose one of the neighbors;
  - copy its state;
  - for each bits in the genome replace the bit with its opposite with probability ω;
- Else do nothing.

We have now to describe the structure of the interacting matrix $\boldsymbol{J}$. I shall deal with a very smooth correspondence between phenotype and genotype: two similar genotypes have also similar phenotype. With this assumptions x represents a strain rather than a species. This hypothesis implies the smoothness of $\boldsymbol{J}$.

First of all I introduce the intraspecies competition. Since the individuals with similar genomes are the ones that share the largest quantity of resources, then the competition is stronger the nearer the genotypes. This implies that $J[x, x] < 0$.

The rest of $\boldsymbol{J}$ is arbitrary. For a classification in terms of usual ecological interrelations, one has to consider together $J[x, y]$ and $J[y, x]$. One can have four cases:

$$
\begin{array}{lll}
J[x,y] < 0 & J[y,x] < 0 & \text{competition} \\
J[x,y] > 0 & J[y,x] < 0 & \text{predation or parasitism} \\
J[x,y] < 0 & J[y,x] > 0 & \text{predation or parasitism} \\
J[x,y] > 0 & J[y,x] > 0 & \text{cooperation}
\end{array}
$$

One has two choices for the geometry of the genetic space.

hypercubic distance: The genetic distance between x and y is given by the number of bits that are different, i.e. $d(x, y) = ||x \oplus y||$, where the norm $||x||$ of a Boolean vector x is defined as $||x|| = (1/N) \sum_{i=1}^{N} x_i$.

linear distance: The genetic space is arranged on a line (excluding the 0 and with periodic boundary conditions) rather than on a hypercube. This arrangement is simpler but less biological; [i]

---

[i]An instance of a similar (sub-)space in real organisms is given by a repeated gene (say a tRNA gene): a fraction of its copies can mutate, linearly varying the fitness of the individual

4.4   Mean field approximation

Since the model is rather complex, it is better to start[j] with the mean field approximation, disregarding the spatial structure. This approximation becomes exact in the limit of large diffusion or when the coupling extends on all the lattice. [k]

Let $n(x)$ be the number of organisms with genetic code x, and n_* the number of empty sites. if N is the total number of cells, we have

$$n_* + \sum_x n(x) = N;$$

$$m = \frac{1}{N} \sum_x n(x) = 1 - \frac{n_*}{N};$$

considering the sums extended to all "not-empty" genetic codes, and indicating with m the fraction of non-empty sites (i.e. the population size).

We shall label with a tilde the quantities after the survival step, with a prime after the reproduction step. The evolution of the system will be ruled by the following equations:

$$\tilde{n}(x) = \pi(x,n)n(x); \tag{26}$$

$$n'(x) = \tilde{n}(x) + \frac{\tilde{n}_*}{N} \sum_y W(x,y)\tilde{n}(y).$$

The matrix $W(x,y)$ is the probability of mutating from genotype y to x. For a given codon mutation probability μ, $W(x,y)$ is given by

hypercubic distance:

$$W(x,y) = \mu^{d(x,y)}(1-\mu)^{l-d(x,y)}; \tag{27}$$

with the "chemical composition" of the gene. [21] This degenerate case has been widely studied (see for instance Alves and Fontanari (1996) [22]). The linear space is equivalent to the hypercubic space if the phenotype $g(x)$ does not depend on the disposition of bits in x (i.e. it depends only on the number of ones – a totalistic function): one should introduce in this case the multiplicity of a degenerate state, which can be approximated to a Gaussian, but if one works in the neighborhood of its maximum (the most common chemical composition) the multiplicity factors are nearly constants. Another example is given by the level of catalytic activity of a protein. A linear space has also been used for modeling the evolution of RNA viruses on HeLa cultures. [23]

[j] We also stop in this lecture with this approximation.

[k] Since the original model is still in development, I shall not study the exact mean field version of the above description, but a simplified one.

linear distance:

$$W(x,y) = \mu \qquad \text{if } |x - y| = 1;$$
$$W(x,x) = 1 - 2\mu; \qquad\qquad\qquad (28)$$
$$W(x,y) = 0 \qquad \text{otherwise.}$$

In any case the mutations represent a diffusion propcess in genic space, thus the matrix W conserves the total population, and

$$\sum_x W(x,y) = \sum_y W(x,y) = 1. \qquad (29)$$

Summing over the genomes x, we can rewrite Eq. (26) for m as

$$\tilde{m} = \frac{1}{N} \sum_x \tilde{n}(x) = \frac{1}{N} \sum_x \pi(x,n)n(x)$$

$$m' = \frac{1}{N} \sum_x n'(x) = \tilde{m} + \frac{\tilde{n}_*}{N^2} \sum_{xy} W(x,y)\tilde{n}(y).$$

Introducing the probability distribution of occupied sites $p(x) = \frac{n(x)}{mN}$ ($\sum_x p(x) = 1$), and the average fitness $\overline{\pi}$ as

$$\overline{\pi} \equiv \frac{1}{mN} \sum_x \pi(x,n)n(x) = \sum_x \pi(x)p(x),$$

and thus

$$\tilde{m} = m\overline{\pi}; \qquad \frac{\tilde{n}_*}{N} = 1 - m\overline{\pi}.$$

Using the property (29) we obtain

$$m' = m\overline{\pi}(2 - m\overline{\pi}). \qquad (30)$$

The stationary condition ($m' = m$) gives

$$1 = \overline{\pi}(2 - m\overline{\pi});$$
$$m = \frac{2\overline{\pi} - 1}{\overline{\pi}^2}.$$

The normalized evolution equation for $p(x)$ is thus

$$p'(x) = \frac{\pi(x,p,m)p(x) + (1 - m\overline{\pi}) \sum_y W(x,y)\pi(x,p,m)p(y)}{\overline{\pi}(2 - m\overline{\pi})}; \qquad (31)$$

where the usual fitness function, Eq.(22), has to be written in term of $p(x)$

$$H(x,p,m) = h[x] + m \sum_y J[x,y]p(y)$$

Notice that Eq.(30) corresponds to the usual logistic equation for population dynamics if we keep the average fitness $\bar{\pi}$ constant.

4.5 Speciation due to the competition among strains

One of the feature of the model is the appearance of the species intended as a cluster of strains connected by mutations. [24]

One can consider the following analogy with a Turing mechanism for chemical pattern formation. The main ingredients are an autocatalytic reaction process (reproduction) with slow diffusion (mutations) coupled with the emission of a short-lived, fast-diffusing inhibitor (competition). In this way a local high concentration of autocatalytic reactants inhibits the growth in its neighborhood, acting as a local negative interaction.

In genetic space, the local coupling is given by the competition among genetically kin individuals. For instance, assuming a certain distribution of some resource (such as some essential metabolic component for a bacterial population), then the more genetically similar two individuals are, the wider the fraction of shared resources is. The effects of competition on strain x by strain y are modeled by a term proportional to the relative abundance of the latter, $p(y)$, modulated by a function that decreases with the phenotypic distance between x and y. Another example of this kind of competition can be found in the immune response in mammals. Since the immune response has a certain degree of specificity, a viral strain x can suffer from the response triggered by strain y if they are sufficiently near in an appropriate phenotypic subspace. Again, one can think that this effective competition can be modeled by a term, proportional to the relative abundance of the strain that originated the response, which decreases with the phenotypic distance.

Let us start with a one dimensional "chemical" model of cells that reproduce asexually and slowly diffuse (in real space), $p = p(x,t)$ being their relative abundance at position x and at time t. These cells constitutively emit a short-lived, fast-diffusing mitosys inhibitor $q = q(x,t)$. This inhibitor may be simply identified with some waste or with the consumption of a local resource (say oxygen). The diffusion of the inhibitor is modeled as

$$\frac{\partial q}{\partial t} = k_0 p + D \frac{\partial^2 q}{\partial x^2} - k_1 q, \tag{32}$$

where k_0, k_1 and D are the production, annihilation and diffusion rates of q. The evolution of the distribution p is given by

$$\frac{\partial p}{\partial t} = \left(A(x,t) - \overline{A}(t)\right) p + \mu \frac{\partial^2 p}{\partial x^2}, \tag{33}$$

$$\overline{A}(t) = \int A(y,t)p(y,t)\mathrm{d}y. \tag{34}$$

The growth rate A can be expressed in terms of the fitness H as

$$A(x,t) = \exp\left(H(x,t)\right). \tag{35}$$

Due to the form of equation (33), at each time step we have

$$\sum_x p(x,t) = 1. \tag{36}$$

The diffusion rate of q, D, is assumed to be much larger than μ. The growth rate A, can be decomposed in two factors, $A(x,t) = A_0(x)A_1(q(x,t))$, where A_0 gives the reproductive rate in absence of q, so $A_1(0) = 1$. In presence of a large concentration of the inhibitor q the reproduction stops, so $A_1(\infty) = 0$. A possible choice is

$$A(x,t) = \exp(H_0(x) - q(x,t)).$$

For instance, $H_0(x)$ could model the sources of food or, for algae culture, the distribution of light.

Since we assumed a strong separation in time scales, we look for a stationary distribution $\tilde{q}(x,t)$ of the inhibitor (Eq. (32)) by keeping p fixed. This is given by a convolution of the distribution p:

$$\tilde{q}(x,t) = J \int \exp\left(-\frac{|x-y|}{R}\right) p(y,t)\mathrm{d}y,$$

where J and R depend on the parameters k_0, k_1, D. In the following we shall use J and R as control parameters, disregarding their origin.

We can generalize this scenario to non-linear diffusion processes of the inhibitor by using the reaction-diffusion equation Eq. (33), with the fitness H and the kernel K given by

$$H(x,t) = H_0(x) - J \int K\left(\frac{x-y}{R}\right) p(y,t)\mathrm{d}y \tag{37}$$

$$K(r) = \exp\left(-\frac{|r|^\alpha}{\alpha}\right), \tag{38}$$

i.e. a symmetric decreasing function of r with $K(0) = 1$. The parameters J and α control the intensity of the competition and the steepness of the interaction, respectively.

Let us consider the correspondence with the genetic space: the quantity x now identifies a genome, the diffusion rate μ is given by mutations, and the inhibitor q (which is no more a real substance) represents the competition among phenotypically related strains. The effects of competition are much faster than the genetic drift (mutations), so that the previous hypotheses are valid. While the competition interaction kernel $K(r)$ is not given by a diffusion process, its general form should be similar to that of Eq. (38): a decreasing function of the phenotypic distance between two strains. We shall refer to the p-independent contribution to the fitness, $H_0[x]$, as the static fitness landscape.

Our model is thus defined by Eqs. (33–38). We are interested in its asymptotic behavior in the limit $\mu \to 0$. Actually, the mutation mechanism is needed only to define the genetic distance and to allow population of an eventual niche. The results should not change qualitatively if one includes more realistic mutation mechanisms.

Let us first examine the behavior of Eq. (33) in absence of competition ($J = 0$) for a smooth static landscape and a vanishing mutation rate. This corresponds to the Eigen model in one dimension: since it does not exhibit any phase transition, the asymptotic distribution is unique. The asymptotic distribution is given by one delta function peaked around the global maximum of the static landscape, or more delta functions (coexistence) if the global maxima are degenerate. The effect of a small mutation rate is simply that of broadening the distribution from a delta peak to a bell-shaped curve[25].

While the degeneracy of maxima of the static fitness landscape is a very particular condition, we shall show in the following that in presence of competition this is a generic case. For illustration, we report in Fig. 17 the numerical computation of the asymptotic behavior of the model for a possible evolutive scenario that leads to the coexistence of three species. We have chosen a smooth static fitness H_0 (see Eq. (48)) and a Gaussian ($\alpha = 2$) competition kernel, using the genetic distance in place of the phenotypic one. The effective fitness H is almost degenerate (here $\mu > 0$ and the competition effect extends on the neighborhood of the maxima), and this leads to the coexistence.

In the following we shall assume again that the phenotypic space is the same of the genotypic space.

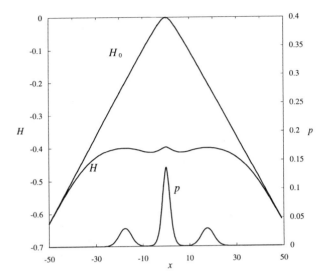

Figure 17: Static fitness H_0, effective fitness H, and asymptotic distribution p numerically computed for the following values of parameters: $\alpha = 2$, $\mu = 0.01$, $H_0 = 1.0$, $b = 0.04$, $J = 0.6$, $R = 10$ and $r = 3$.

Evolution near a maximum

We need the expression of p if a given static fitness $H(x)$ or a static growth rate $A(x)$ has a smooth, isolated maximum for $x = 0$ (*smooth maximum* approximation). Let us assume that

$$A(x) \simeq A_0(1 - ax^2), \tag{39}$$

where $A_0 = A(0)$.

The generic evolution equation (master equation) for the probability distribution is

$$\alpha(t)p(x, t+1) = \left(1 + \mu\frac{\delta^2}{\delta x^2}\right) A\big(x, p(t)\big)p(x, t); \tag{40}$$

where the discrete second derivative $\delta^2/\delta x^2$ is defined as

$$\frac{\delta f(x)}{\delta x^2} = f(x+1) + f(x-1) - 2f(x),$$

and $\alpha(t)$ maintains the normalization of $p(t)$. In the following we shall mix freely the continuous and discrete formulations of the problem.

Summing over x in Eq. (40) and using the normalization condition, Eq. (36), we have:

$$\alpha = \sum_x A(x, \boldsymbol{p})p(x) = \overline{A}. \qquad (41)$$

The normalization factor α thus corresponds to the average fitness. The quantities A and α are defined up to an arbitrary constant.

If A is sufficiently smooth (including the dependence on $\boldsymbol{p}$), one can rewrite Eq. (40) in the asymptotic limit, using a continuous approximation for x as

$$\alpha p = Ap + \mu \frac{\partial^2}{\partial x^2}(Ap), \qquad (42)$$

Where we have neglected to indicate the genotype index x and the explicit dependence on $\boldsymbol{p}$. Eq. (42) has the form of a nonlinear diffusion-reaction equation. Since we want to investigate the phenomenon of species formation, we look for an asymptotic distribution $\boldsymbol{p}$ formed by a superposition of several non-overlapping bell-shaped curves, where the term non-overlapping means almost uncoupled by mutations. Let us number these curves using the index i, and denote each of them as $p_i(x)$, with $p(x) = \sum_i p_i(x)$. Each $p_i(x)$ is centered around $\overline{x}_i$ and its weight is $\int p_i(x)dx = \gamma_i$, with $\sum_i \gamma_i = 1$. We further assume that each $p_i(x)$ obeys the same asymptotic condition, Eq. (42) (this is a sufficient but not necessary condition). Defining

$$\overline{A}_i = \frac{1}{\gamma_i} \int A(x)p_i(x)dx = \alpha, \qquad (43)$$

we see that in a stable ecosystem all quasi-species have the same average fitness.

Substituting $q = Ap$ in Eq. (42) we have (neglecting to indicate the genotype index x, and using primes to denote differentiation with respect to it):

$$\frac{\alpha}{A}q = q + \mu q''.$$

Looking for $q = \exp(w)$,

$$\frac{\alpha}{A} = 1 + \mu(w'^2 + w''),$$

and approximating $A^{-1} = A_0^{-1}\left(1 + ax^2\right)$, we have

$$\frac{\alpha}{A_0}(1 + ax^2) = 1 + \mu(w'^2 + w''). \qquad (44)$$

A possible solution is

$$w(x) = -\frac{x^2}{2\sigma^2}.$$

Substituting into Eq. (44) we finally get

$$\frac{\alpha}{A_0} = \frac{2 + a\mu - \sqrt{4a\mu + a^2\mu^2}}{2}. \tag{45}$$

Since $\alpha = \overline{A}$, α/A_0 is less than one we have chosen the minus sign. In the limit $a\mu \to 0$ (small mutation rate and smooth maximum), we have

$$\frac{\alpha}{A_0} \simeq 1 - \sqrt{a\mu}$$

and

$$\sigma^2 \simeq \sqrt{\frac{\mu}{a}}. \tag{46}$$

The asymptotic solution is

$$p(x) = \gamma \frac{1 + ax^2}{\sqrt{2\pi}\sigma(1 + a\sigma^2)} \exp\left(-\frac{x^2}{2\sigma^2}\right),$$

so that $\int p(x)dx = \gamma$. The solution is a bell-shaped curve, its width σ being determined by the combined effects of the curvature a of maximum and the mutation rate μ.. In the next section, we shall apply these results to a quasi-species i. In this case one should substitute $p \to p_i$, $\gamma \to \gamma_i$ and $x \to x - \overline{x}_i$.

For completeness, we study also the case of a *sharp maximum*, for which $A(x)$ varies considerably with x. In this case the growth rate of less fit strains has a large contribution from the mutations of fittest strains, while the reverse flow is negligible, thus

$$p(x - 1)A(x - 1) \gg p(x)A(x) \gg p(x + 1)A(x + 1)$$

neglecting last term, and substituting $q(x) = A(x)p(x)$ in Eq. (40) we get:

$$q(x) = \begin{cases} \frac{\alpha}{A_0} = 1 - 2\mu & \text{for } x = 0 \\ \frac{\mu}{(\alpha A(x) - 1 + 2\mu)}q(x - 1) & \text{for } x > 0 \end{cases} \tag{47}$$

Near $x = 0$, combining Eq. (47), Eq. (47)and Eq. (39)), we have

$$q(x) = \frac{\mu}{(1 - 2\mu)ax^2}q(x - 1).$$

In this approximation the solution is

$$q(x) = \left(\frac{\mu}{1 - 2\mu a}\right)^x \frac{1}{(x!)^2},$$

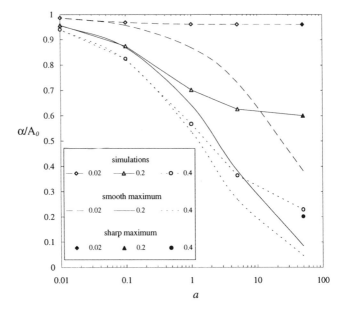

Figure 18: Average fitness α/A_0 versus the coefficient a, of the fitness function, Eq. (39), for some values of the mutation rate μ. Legend: *numerical resolution* corresponds to the numerical solution of Eq. (40), *smooth maximum* refers to Eq. (45) and *sharp maximum* to Eq. (47)

and

$$y(x) = A(x)q(x) \simeq \frac{1}{A_0}(1 + ax^2) \left(\frac{\mu A_0}{\alpha a}\right)^x \frac{1}{x!^2}.$$

One can check the validity of these approximations by solving numerically Eq. (40); the comparisons are shown in Fig. 18. One can check that the *smooth maximum* approximation agrees with the numerics for for small values of a, when $A(x)$ varies slowly with x, while the *sharp maximum* approximation agrees with the numerical results for large values of a, when small variations of x correspond to large variations of $A(x)$.

Speciation

I shall now derive the conditions for the coexistence of multiple species. Let us assume that the asymptotic distribution is formed by L delta peaks p_k, $k = 0, \ldots, L - 1$, for a vanishing mutation rate (or L non-overlapping bell shaped curves for a small mutation rate) centered at y_k. The weight of each quasi species is γ_k, i.e.

$$\int p_k(x)dx = \gamma_k, \qquad \sum_{k=0}^{L-1} \gamma_k = 1.$$

The quasi-species are ordered such as $\gamma_0 \geq \gamma_1, \ldots, \geq \gamma_{L-1}$.
The evolution equations for the p_k are ($\mu \to 0$)

$$\frac{\partial p_k}{\partial t} = (A(y_k) - \overline{A})p_k,$$

where $A(x) = \exp(H(x))$ and

$$H(x) = H_0(x) - J \sum_{j=0}^{L-1} K\left(\frac{x-y_j}{R}\right)\gamma_j.$$

The stability condition of the asymptotic distribution is $(A(y_k)-\overline{A})p_k = 0$, i.e. either $A(y_k) = \overline{A} = $ const (degeneracy of maxima) or $p_k = 0$ (all other points). In other terms one can say that in a stable environment the fitness of all individuals is the same, independently on the species.

The position y_k and the weight γ_k of the quasi-species are given by $A(y_k) = \overline{A} = $ const and $\partial A(x)/\partial x\big|_{y_k} = 0$, or, in terms of the fitness H, by

$$H_0(y_k) - J \sum_{j=0}^{L-1} K\left(\frac{y_k - y_j}{R}\right)\gamma_j = \text{const}$$

$$H_0'(y_k) - \frac{J}{R}\sum_{j=0}^{L-1} K'\left(\frac{y_k - y_j}{R}\right)\gamma_j = 0$$

Let us compute the phase boundary for coexistence of three species for two kinds of kernels: the exponential (diffusion) one ($\alpha = 1$) and a Gaussian one ($\alpha = 2$).

I assume that the static fitness $H_0(x)$ is a symmetric linear decreasing function except in the vicinity of $x = 0$, where it has a quadratic maximum:

$$H_0(x) = b\left(1 - \frac{|x|}{r} - \frac{1}{1+|x|/r}\right) \tag{48}$$

so that close to $x = 0$ one has $H_0(x) \simeq -bx^2/r^2$ and for $x \to \infty$, $H_0(x) \simeq b(1 - |x|/r)$. Numerical simulations show that the results are qualitatively independent on the exact form of the static fitness, providing that it is a smooth decreasing function.

Due to the symmetries of the problem, we have one quasi-species at $x = 0$ and two symmetric quasi-species at $x = \pm y$. Neglecting the mutual influence of the two marginal quasi-species, and considering that $H_0'(0) = K'(0) = 0$, $K'(y/R) = -K'(-y/r)$, $K(0) = J$ and that the three-species threshold is given by $\gamma_0 = 1$ and $\gamma_1 = 0$, we have

$$\tilde{b}\left(1 - \frac{\tilde{y}}{\tilde{r}}\right) - K(\tilde{y}) = -1,$$

$$\frac{\tilde{b}}{\tilde{r}} + K'(\tilde{y}) = 0.$$

where $\tilde{y} = y/R$, $\tilde{r} = r/R$ and $\tilde{b} = b/J$. I introduce the parameter $G = \tilde{r}/\tilde{b} = (J/R)/(b/r)$, that is the ratio of two quantities, one related to the strength of inter-species interactions (J/R) and the other to intra-species ones (b/r). In the following I shall drop the tildes for convenience. Thus

$$r - z - G\exp\left(-\frac{z^\alpha}{\alpha}\right) = -G,$$

$$Gz^{\alpha-1}\exp\left(-\frac{z^\alpha}{\alpha}\right) = 1,$$

For $\alpha = 1$ we have the coexistence condition

$$\ln(G) = r - 1 + G.$$

The only parameters that satisfy these equations are $G = 1$ and $r = 0$, i.e. a flat landscape $(b = 0)$ with infinite range interaction $(R = \infty)$. Since the coexistence region reduces to a single point, it is suggested that $\alpha = 1$ is a marginal case. Thus for less steep potentials, such as power law decrease, the coexistence condition is supposed not to be fulfilled.

For $\alpha = 2$ the coexistence condition is given by

$$z^2 - (G + r)z + 1 = 0,$$

$$Gz\exp\left(-\frac{z^2}{2}\right) = 1.$$

One can solve numerically this system and obtain the boundary $G_c(r)$ for the coexistence. In the limit $r \to 0$ (static fitness almost flat) one has

$$G_c(r) \simeq G_c(0) - r \tag{49}$$

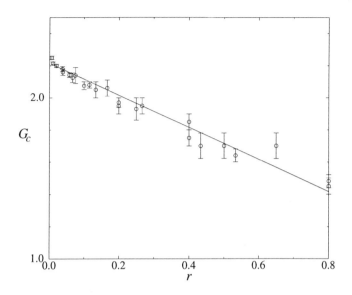

Figure 19: Three-species coexistence boundary G_c for $\alpha = 2$. The continuous line represents the analytical approximation, Eq. (49), the circles are obtained from numerical simulations. The error bars represent the maximum error.

with $G_c(0) = 2.216\ldots$. Thus for $G > G_c(r)$ we have coexistence of three or more quasi-species, while for $G < G_c(r)$ only the fittest one survives.

I have solved numerically Eqs. (33–38) for several different values of the parameter G, considering a discrete genetic space, with N points, and a simple Euler algorithm. The results, presented in Fig. 2, are not strongly affected by the integration step. The error bars are due to the discreteness of the changing parameter G. The boundary of the multi-species phase is well approximated by Eq. (49); in particular, I have checked that this boundary does not depends on the mutation rate μ, at least for $\mu < 0.1$, which can be considered a very high mutation rate for real organisms. The most important effect of μ is the broadening of quasi-species curves, which can eventually merge.

4.6 Final considerations

Possible measures on this abstract ecosystem model concern the origin of complexity (for instance considered equivalent to the entropy) in biological systems. One expects that steady ecosystems reach an asymptotic value of the complexity that maximizes the explotation of energy, while perturbed ecosys-

44

tems exhibit lower complexity. Another interesting topic regards the behavior
of complexity in relation with the spatial extension of a system.

References

1. A. Burks (ed.) *Essays on Cellular Automata* (Univ. of Illinois Press, 1970).
2. K. Sigmund, *Games of Life* (Oxford University Press, New York 1993).
3. E. Belekamp, J. Conway and R. Guy, editors, *Winning Ways*, vol. 2 *What is Life?* (Academic Press, New York, 1982), ch. 25; M. Gardner, *Weels, Life and Other Stories* (Freeman, New York, 1983), ch. 20-22.
4. AA. VV. *Physica* **10D** (North Holland, 1984).
5. S. Wolfram, editor, *Theory and Applications of cellular Automata* (World Scientific, Singapore 1987).
6. P. Grassberger, J. Stat. Phys **45**, 27 (1986) and Physica **10D**, 52 (1984).
7. F. Bagnoli, R. Rechtman, S. Ruffo, Phys. Lett. A **172**, 34 (1992).
8. A. Georges and P. le Doussal, J. Stat. Phys. **54**, 1011 (1989).
9. E. Domany and W. Kinzel, Phys. Rev. Lett. **53**, 311 (1984).
10. P. Bak, C. Tang and K. Wiesenfeld, Phys. Rev. Lett. **59**, 381 (1987).
11. B. Chopard and M. Droz, *Cellular Automata and Modeling of Complex Physical Systems*, Springer proceedings in Physycs **46**, P. Manneville et al. , editors (Springer, Berlin 1989) p. 130.
12. F. Bagnoli, P. Palmerini and R. Rechtman, *Mapping Criticality into Self Criticality*, Phys. Rev. E. **55**, 3970 (1997).
13. Wegener *The Complexity of Boolean Functions* (Wiley, New York 1987).
14. F. Bagnoli, International Journal of Modern Physics C **3**, 307 (1992).
15. H.A. Gutowitz, D. Victor and B.W. Knight, Physica **28D**, 28 (1987).
16. F. Bagnoli, R. Rechtman, S. Ruffo, Physica A **171**, 249 (1994).
17. P. Gibbs and D. Stauffer, *Search for Asymptotic Death in Game of Life*, Int. J. Mod. Phys. C. **8**, 601 (1997).
18. D. Stauffer, private communication (1997).
19. P. Bak, K. Chen and M. Creutz, Nature **342**, 780 (december 1989).
20. F. Bagnoli, J. Stat. Phys. **86**, 151 (1996).
21. F. Bagnoli and P. Lió, J. Theor. Biol. **173**, 271 (1995).
22. D. Alves and J. F. Fontanari, Phys. Rev. E **54**, 4048 (1996).
23. L.S. Tsimring, H. Levine and D.A. Kessler, Phys. Rev. Lett. **76**, 4440 (1996); D. A. Kessler, H. Levine, D. Ridgway and L. Tsmiring, J. Stat. Phys. **87**, 519 (1997).
24. F. Bagnoli and M. Bezzi, *Speciation as Pattern Formation by Competition in a Smooth Fitness Landscape*, Phys. Rev. Lett. (1997, in press).

25. F. Bagnoli and M. Bezzi, *Competition in a Fitness Landscape*, Fourth European Conference on Artificial Life, P. Husbands and I. Harvey (eds.), The MIT Press (Cambridge, Massachussets, 1997) p. 101 and cond-mat/9702134 (1997).

BAROLO: Biological Ageing Research on Long-Lived Organisms

DIETRICH STAUFFER

Institute for Theoretical Physics, Cologne University, 50923 Köln, Germany

The phenomenological ageing theory of Azbel as well as many Monte Carlo simulations based on Penna's mutation accumulation model are reviewed. We emphasize our difficulties with sex: Why did nature invent it; are men useful for anything? These theories apply not only to humans but also to multicellular animals and plant, though not to bacteria, virus, or spin glasses.

1 Introduction

"Sex and the Single Bit" was the originally intended title of this review, following an old Hollywood movie. But such amoralic behavior was not tolerated by the Chief Organizer who instead invented the above acronym, which also makes justified propaganda for the very valuable red wine of the Torino region. Ageing simulations were reviewed by Bernardes[1] and more superficially by this writer,[2] and thus the present article concentrates on the later research, with a phenomenological theory in the next section, followed by Monte Carlo simulations with sexual reproduction. Age is defined here not in terms of beauty or scientific originality but very simply and quantitatively in terms of the mortality, the probability to die within the next year.

2 Azbel theory

The Gompertz law of the 19th century tells us that the mortality $q(x)$ of humans increases exponentially with age x, after childhood troubles have been overcome. Figure 1 shows recent Japanese data indicating roughly

$$q \propto \exp(bx) \tag{1}$$

for both males and females. More precisely, child mortality is very high and does not follow this Gompertz law, and men have a mortality roughly twice as high as women, except at very old age when men finally achieve equal rights.

Traditionally, the mortality q is defined as the fraction at age x which survives until age $x + 1$:

$$q(x) = [N(x) - N(x + 1)]/N(x). \tag{2}$$

Here $N(x)$ is the number of survivors at age x in a stationary population, and the age x is measured in suitable units, like a year for humans and a day for

48

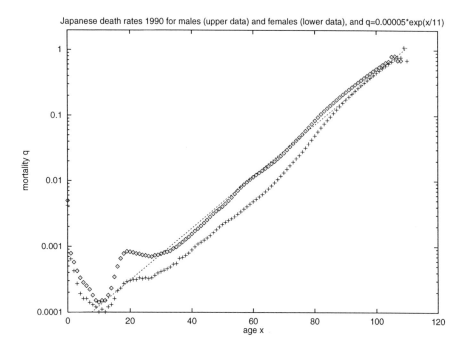

Figure 1: Japanese mortality: diamonds for males, plus signs for females. The straight line shows the Gompertz law with slope $b = 1/(11\ \text{years})$.

fruitflies. Obviously, this definition leads to $q \leq 1$ independent of the choice of the time unit, and it thus contradicts Eq. (1) where q goes to infinity for age x going to infinity. Azbel[3] solves this definition problem by replacing Eq. (2) with

$$q(x) = \ln[N(x)/N(x+1)]. \qquad (3)$$

Now the mortality q can increase beyond unity as required by the Gompertz law (1). (Basically, one should define $q = -d\ln N/dx$; approximating this derivative by a constant in the age interval from x to $x+1$ leads to Eq. (3).) This definition is already used in Figure 1, though q there does not yet significantly surpass unity.

Other than humans, only some mediterranean fruitflies were studied with more than a million individuals.[4] Here, an interesting effect was observed for very old age: The mortality saturated or even decayed for increasing age. If humans would learn such survival techniques from these medflies, a small fraction of us might live thousand years.[3] To explain such and other data, Azbel[3] introduced a distribution of Gompertz factors b, Eq. (1), in the population.

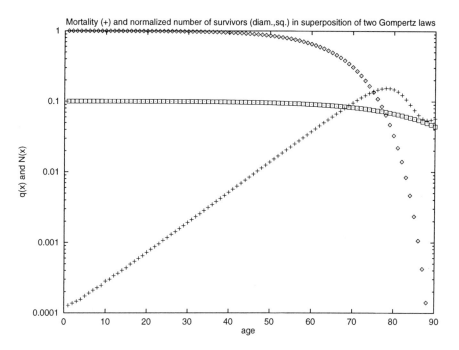

Figure 2: Superposition of two Gompertz laws result in local maximum of total mortality.

This heterogeneity within one species means that some genetically well endowed but rare families live much longer than the vast majority. And this heterogeneity can then explain such deviations from the Gompertz law.

Figure 2 shows a simple example, where the population starts with 10^5 normal individuals having a mortality $0.0001\exp(0.1x)$ and 10^4 privileged individuals having a mortality $0.0001\exp(0.07x)$. The numbers of survivors (normalized by 10^5) of the two groups are shown as squares and diamonds, and the resulting mortality q for the total population as plusses. We see that this addition of two Gompertz laws gives first the normal exponential increase of q with age x. But when the surviving number of privileged people is no longer small compared to the survivors of the original majority, here at an age of about 80, then deviations from the simple Gompertz law are seen, which lead here to a decrease of mortality after that age.

More precisely, Azbel[3] defines a dimensionless mortality $q(x)/b \propto \exp(bx)$ independent of the time unit, a characteristic age X and a proportionality factor A which both are the same for the whole species, while b may differ for different groups:

$$q(x)/b = A \exp(b(x - X)). \qquad (4a)$$

With dimensionless time units $y = bx$, $Y = bX$ and thus $q/b \simeq -d\ln N/dy$ this equation reads

$$-d\ln N/dy = A \exp(y - Y). \qquad (4b)$$

In other words, if the logarithm $\ln(q/b)$ of the dimensionless mortality is plotted versus age x then all straight-line fits for humans go through the same point $\ln A$ at $x = X$, even though their slope b changes drastically with country and century. Human lifetables from Sweden, Japan and Germany of the last two centuries, involving more than 10^9 people and a thousand-fold change in the mortality, follow at intermediate ages roughly this law with $X = 103$ years and A around 11, while $b(1/b \simeq 11$ years in fig.1) changes with improved living conditions and health care. Thus X seems to be determined by our genes, while b is influenced by the environment (if we count all deaths) and also may differ for different individuals. Similar fits were successful for medflies where X is measured in days instead of years. When we discard all deaths due to infections, accidents etc as "premature" deaths and count only those due genetic factors (which we can do easily in the computer model of the next section), then also the factor b is determined completely by our genes.

Instead of one single b a whole distribution of b−values may exist in a population $N(x) = \int n(x,b)\, db$. Then we see from Azbel's law (4a), that for $x < X$ the mortality is higher for small b while for $x > X$ the mortality is higher for large b, provided the exponential in Eq. (4) dominates. Thus first predominantly the groups with small b die out, while later, for age above X, the survivors with large b die out more quickly. In this way, deviations from the simple Gompertz law (1) become possible, and many aspects of the Azbel theory seem to be compatible with data on humans and medflies.[3] The crucial question may be if the $b-$ distribution extends down to $b = 0$ or is restricted to b above some positive limit. If for the oldest old we have $b(x - X) \gg 1$, then these distinctions become important and may explain the qualitative difference between humans and medflies [3,4].

More precisely, if the $b−$distribution extends down to a positive limit $b_{min} > 0$, then for $b_{min}(x-X) \gg 1$ the mortality increases as $\exp(b_{min}(x-X))$. If instead the $b−$distribution extends down to $b = 0$, approaching a constant or a power law in b for small b, then[3] the resulting mortality for the whole population decays for sufficiently large age x as $1/(x-X)$. Other distributions may lead to other laws.

An unsolved problem is how this concept can be reconciled with Darwinistic evolution through selection of the fittest: Those families with the more favorable genetic setup should have more children and thus after sufficiently many generations should dominate in the population, changing an initial heterogeneity into a nearly homogeneous population. Thus effects like Fig.2 would vanish due to evolution, and the b−distribution should become nearly a delta function. This is indeed what Monte Carlo simulations with two groups in the population have given, [5] resulting finally into the simple Gompertz law with no explanation of a mortality decreasing with increasing age for the oldest old.

Another disadvantage of this phenomenological description is of course that the mortality curve $q(x)$ requires the knowledge of the b−distribution in the population, i.e. of another curve $n(x = 0, b)$. In principle, this allows for infinitely many fittable parameters. However, according to Azbel many conclusions do not depend on the details of this curve.

Does the theory allow us to live thousand years? No ! [3] According to Azbel, humans and medflies are described by the same functions (4) but with different parameters: the distribution $n(x = 0, b)$ is too different to allow the medfly effect of mortality decreasing with age to occur with us. Thus we should hurry up to study a more microscopic model, in line with the tradition of physics to look for explanations via basic particles, beyond phenomenological theory. Atoms and quarks then correspond to individuals and their genome.

3   Penna model

In Statistical Physics, energy and entropy counteract each other with the result that neither the energy nor the negative entropy but the free energy is minimal. Similarly, the mutation accumulation hypothesis [6] explains ageing from the counterbalance of Darwinistic selection, which prefers perfect individuals, and random detrimental mutations, which cause disorder and diseases. These mutations are inheritable and can happen with equal probability at any stage in life. Thus at first one may think that they increase the mortality $q(x)$ equally for all ages x. But this is not so once we take into account the effects of reproduction: A dangerous mutation killing us before we have children will die with us; a mutation killing us in old age will be given on to our children and thus may remain forever in the population. Thus in a stationary equilibrium of new mutations versus selection pressure, mutations affecting us at old age will stay in the population much more often than mutations afflicting already young people. Therefore hereditary diseases increase the mortality much more in old than in young age, as required by Eq. (1).

The Penna model [7] of 1994 is now by far the most widely used computer simulation technique to predict $q(x)$ in ageing populations, and to my knowledge the only one reproducing approximately Eq. (1). The maximum age is divided into 32 intervals, called years, on a computer with 32-bit words. In each interval, one bad hereditary disease, caused by an earlier mutation, may start to affect the individual until the end of its life. Three or more such active diseases kill us. Each bit corresponds to one year and is set if and only if starting from the year to which the bit position belongs a dangerous hereditary disease threatens our life. The human genome has 10^5 and not just 32 genes, but most of these genes do not cause death if mutated. Besides, qualitatively there was not much difference between 32, 64 and 128 bits taken into account. After a minimum age of reproduction, typically 8 years, each adult individual gets typically one child per year until it dies. This child can differ at one randomly selected bit position from the genome of the parent: If the parent bit was not yet set the child bit is now set; if the parent bit was already set before, the child bit remains set (or is set back to zero). Because of this randomness in these mutations, the deterministic "mutational meltdown"[8] is avoided which would eventually lead to an extinction of the whole population. Besides these genetic deaths due to bad mutations, there are also deaths due to restrictions of food and space which are taken into account by an additional death probability N/N_{max} (Verhulst factor), where N is the total population and N_{max} a maximum population.

Figure 3 from Meisgen[9] shows that this model gives a roughly exponential increase of mortality q with age x, as required by Eq. (1). However, in order to find this we had to subtract the "premature"[3] deaths due to the Verhulst factor from the genetically determined deaths due to mutations. More easily, one can do this approximately after the simulation by the Oliveira trick:[10] Subtract from the total mortality that at very young age since the latter is hardly due to mutations. Thus the diamonds in Figure 3 represent the total mortality, equal to about 0.145 at young age from the Verhulst factor, and the plusses are the same data with this number subtracted. In rich countries at peace, death by hunger and infectious diseases should be relatively rare (the Verhulst factor being replaced by birth control), and then the plusses should correspond to reality except for traffic accidents etc.

This model was, after the reviews of Bernardes[1] and Stauffer[2], applied to the age distribution in West Germany,[11] to the accidental extinction of very small populations,[12] the effects of hunting on the social life of wolves,[13] and (based on a Bagnoli idea during this conference) age-dependent mutation rates and child mortality,[14] and finally sexual cannibalism.[15] More recently, exact solutions,[16] confirmation of Azbel universality,[17] power law distribution of family sizes,[18] sex hope for old men,[19] and chaos[20] were added. Chapter 2 of *Sex, Money, War, and Computers*[21] gives a more complete review.

An important aspect of asexual reproduction which is kept also by its

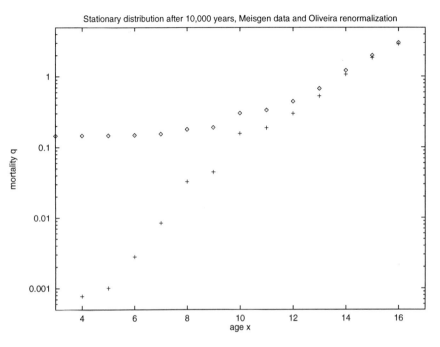

Figure 3: Asexual Penna-model results with total mortality (diamonds) and only genetic mortality (plus signs); the latter one excludes death due to the age-independent Verhulst factor (lack of food and space).

sexual variant[22] is that after some time all survivors are offspring of the same mother and father. Thus they all may share bits set at the same years in youth, causing fluctuations which do not go away if we simulate over longer time. This biblical effect also causes problems in parallel computing[9] since after some time all survivors of asexual reproduction are located on the same processor if no load balancing is made.

4 Generalization to sex

Why has mother nature invented sex ? Bacteria can life (mostly) without it but most of the more complicated organisms enjoy sex, presumably since hundreds of million years. Typical arguments are that the greater variety due to sexual recombination gives an evolutionary advantage similar to mathematical optimization via genetic algorithms. But how can the short-term disadvantage, that (with exceptions) men do not get pregnant, dont do the dishes and watch football games, be counterbalanced by advantages visible only after thousands of years ? We have here again the problem that evolution is supposed to tranform the first airplane of the Wright brothers to a Boeing 747 without landing in between.

The dangers of marriage are well publicized. Van Voorhies[23] warned already, that mated males live shorter than unmated ones. And Clark[24] wrote a book on sex and the origin of death. (Never mind that his thesis is more that not sex but multicellularity and the ensuing distinction between soma and germ-line cells causes the death of the soma, i.e. of us.) Moreover, some women,[25,26] as well as some married men apparently under pressure from their wives,[27,28] have made computer simulations warning that sex endangers a species and that nature would be better off without men.

The first of these simulations by Redfield[25] was computationally very attractive since an infinitely large population was studied by transforming small arrays of survival probabilities as a function of the number of mutations. (No age structure was taken into account). Starting from one distribution of mutations in the offspring, each survived with probability 0.9^k to adulthood, where k is the number of mutations in its genome. Then new mutations were added randomly via a Poisson distribution, with mutation rates which can be different for males and females. Then children get a genome with a number k of mutations which on average equals half of the sum of the mutations of mother and father; more precisely, the mutations for the children followed a binomial distribution corresponding to random crossover between the mother's and the father's genome. This procedure constitutes one iteration, and after 50 to 100 iterations a stationary state is reached, within seconds on a workstation.

With the same mutation rate for females as for males, this model gives the same average survival rates whether sexual or asexual reproduction is used, and that alone shows then men are dangerous since they eat away the food from the women without producing any advantage in the survival rates. When, as observed biologically, the male mutation rate is higher than the female one, then sex *reduces* the survival rates compared to parthenogenetic (asexual) reproduction by females only.

However, the above model treats all mutations as "co-dominant"[25] whereas in reality most mutations are recessive and only a small fraction is dominant. Recessive mutations affect us only if both father and mother had them, while dominant mutations threaten our life already when one of the two parents had them. Bernardes[1] already emphasized the distinction between recessive and dominant mutations, and indeed when this distinction was made in the Redfield model, the introduction of sex[10] drastically *increased* instead of decreased the survival rates compared with asexual reproduction. The Redfield simulations then corresponded to extreme inbreeding where all mutations exist in both parents, and inbreeding indeed is dangerous.

No age was included in the Redfield model, but when sexual reproduction was combined with the Penna model it gave reasonable results. [10] The model then explained why women survive after menopause even though in Pacific Salmon, and the correponding asexual simulations, [29] life stops after the end of reproduction. Adding somatic mutations to the hereditary mutations discussed above (an effect also emphasized by Clark [24]), and assuming that men have also a higher somatic mutation rate than women, the difference between male and female mortality at intermediate age and their convergence at high age (Fig. 1) could be explained. [30] (Alternatively, this difference was explained by a higher female resistance to disease. [31]) Thus sex seemed justified, and not an error of nature, at this stage.

Unfortunately, Bernardes [27] pointed out that "parthenogenetic meiosis" is about as good for the survival of a species as is sexual recombination while it avoids hungry unproductive males. In this asexual form of reproduction, the crossover between the two DNA strands (the genome bits) happens within each individual. Dasgupta [28] looked at a similar model by asking specifically for the advantage of crossover: pure cloning (one strand of DNA copied) versus meiotic parthenogenesis (crossover between two DNA strands within one individual without any recombination from different individuals). No clear advantages of one versus the other were found, even after a catastrophic change like a meteor killing the dinosaurs. These surprising results were confirmed by independent simulations. [26] Thus again sex seems to be an error of nature, and women would be better off without men. The latter ones could feel challenged to improve the models to justify their own existence.

5 Discussion

This paper presented one phenomenological and one more microscopic description of biological ageing. These genetic approaches seem to be what most recent papers emphasize. That does not prove, however, that this approach is correct. Perhaps instead we get old because of wear and tear in faculty meetings, similar to the bodies of top athletes. Or somatic mutations, which we used here more as a side effect to explain why women live longer than men, are the dominating factor. [24] Or programmed cell death, which surely is crucial for the development of the embryo, also tells our heart to stop beating after 10^2 years. The observation that average life span corresponds roughly to about ten oxygen molecules consumed per body atom of an animal [32] suggests a rather common cause for death. But the different lifestyles of Pacific Salmon [29] and of California Redwood Trees[33] indicate that not all living beings are the same, even if both extremes have been described by the Penna model. Besides, bacteria have shown a way to immortality, provided we do not define cell division as death.

Presently unsolved problems are nature's preference for sex compared with meiotic parthenogenesis, the coevolution of different species beyond simple prey-predator pairs, ageing in geographically different but connected spaces, the influence of randomness[34] (as in thermal fluctuations) on ageing, child mortality[3] beyond the simple simulation of Vandevalle[14] or rescaling of the time[35] etc. I hope to see soon more papers about these questions.

Will future simulations cure ageing ? Presumably not, but understanding nature may give hints for improvement. Similarly, mankind's problem with energy supply was not solved when conservation of mechanical energy was proven by Newton's laws, or its equivalence to heat understood. But understanding the Carnot cycle in thermodynamics lead to an understanding of steam engine efficiencies and their improvement. Presently, we are not yet at Newton's stage in ageing theories, but rather behave like the alchemists did science hundreds of years ago. On their basis, modern chemistry developped 200 years ago, while the correct quantum chemistry came much later.

Acknowledgments

We thank Debashish Chowdhury for a full professorial reading of the manuscript and M. Azbel and N. Jan for helpful comments.

References

1. A.T. Bernardes, Monte Carlo Simulations of Biological Ageing, page 359 in: *Annual Reviews of Computational Physics*, vol. IV, edited by D. Stauffer, World Scientific, Singapore 1996;.
2. D. Stauffer, Computers in Physics **10**, 341 (1996).
3. M. Ya. Azbel, Proc. Roy. Soc. London B **263**, 1449 (1996), Phys. Repts. **288**, 545 (1997) and Physica A, in press.
4. J.R. Carey, P. Liedo, D. Orozco, J.W. Vaupel, Science **258**, 457 (1992).
5. S. Moss de Oliveira, P.M.C. de Oliveira and D. Stauffer, Physica A **221**, 453 (1995).
6. L. Partridge and N. H. Barton, Nature **362**, 305 (1993); B. Charlesworth, *Evolution in Age-Structured Populations*, second edition (Cambridge University Press, Cambridge 1994); M.Rose, *Evolutionary Biology of Aging* (Oxford University Press, New York 1991);.
7. T.J.P. Penna, J. Stat. Phys. **78**, 1629 (1995).

8. M. Lynch and W. Gabriel, Evolution **44**, 1725 (1990).

9. F. Meisgen, Int. J. Mod. Phys. C **8**, 575 (1997).

10. D. Stauffer, P.M.C. de Oliveira, S. Moss de Oliveira and R.M. Zorzenon dos Santos, Physica A **231**, 504 (1996).

11. T.J.P. Penna and D. Stauffer, Z. Physik B **101**, 469 (1996).

12. K.F. Pal, Int. J. Mod. Phys. C **7**, 899 (1996).

13. S.J. Feingold, Physica A **231**, 499 (1996); S. Cebrat and J. Kakol, Int.J.Mod.Phys. C **8**, 417 (1997); D. Makowiec, Physica A **245**, 99 (1997).

14. K.N. Berntsen, Int. J. Mod. Phys. C **7**, 731 (1996).

15. N. Vandewalle, Physica A **245**, 113 (1997).

16. A.F.R. Toledo Piza, Physica A **242**, 195 (1997); R.M.C. de Almeida, S. Moss de Oliveira and T.J.P Penna, preprint.

17. A. Racco, M. Argollo de Menezes and T.J.P. Penna, preprint for Theory in Biosciences.

18. G.A. Medeiros and S. Moss de Oliveira, priv. comm.

19. S.G.F. Martins and T.J.P. Penna, preprint.

20. A.T. Bernardes, J.G. Moreira and A. Castro-e-Silva, European Physical Journal B **1**, in press (1998).

21. S. Moss de Oliveira, P.M.C de Oliveira and D. Stauffer, *Sex, Money, War, and Computers*, submitted to Lecture Notes in Computational Science and Engineering (Springer Verlag, Berlin and Heidelberg).

22. P.M.C. de Oliveira, S. Moss de Oliveira and D. Stauffer, Theory in Biosciences **116**, 65 (1997). See also Y.C. Zhang, M. Serva and M. Polykarpov, J. Stat. Phys. **58**, 849 (1990).

23. W.A. van Voorhies, Nature **360**, 456 (1992).

24. W.R. Clark, *Sex and the Origins of Death* (Oxford University Press, New York 1996).

25. R.J. Redfield, Nature **369**, 145 (1994).

26. J.S. de Sa Martins and S. Moss de Oliveira, priv.comm.

27. A.T. Bernardes, J. Stat. Phys. 86, 431 (1997) and Ann.Physik **5**, 539 (1996).

28. S. Dasgupta, Int. J. Mod. Phys. C **8**, 605 (1997).

29. T.J.P. Penna, S. Moss de Oliveira and D. Stauffer, Phys. Rev. E **52**, R3309 (1995).

30. S. Moss de Oliveira, P.M.C. de Oliveira and D. Stauffer, Braz. J. Phys. **26**, 626 (1996).

31. T.J.P. Penna and D. Wolf, Theory in Biosciences **116**, 118 (1997).

32. M.Ya. Azbel, Proc. Natl. Acad. Sci. USA **91**, 12453 (1994).

33. M. Argolo de Menezes, A. Racco, and T.J.P. Penna, Physica A **233**, 221

(1996).

34. J. Thoms, P. Donahue and N. Jan, J. Physique I **5**, 935 (1995); J. Thoms, P. Donahue, D.L. Hunter, and N. Jan, J. Physique I **5**, 1689 (1995).

35. A. Strotmann, priv.comm., Jan. 1997.

BACTERIAL TRANSLATION MODELING AND STATISTICAL MECHANICS

GIOVANNA GUASTI

Institute for Theoretical Physics and Astrophysics
Universität Potsdam PF 601553
D-14415 Potsdam, Germany

1 Introduction

We are interested in modeling the evolution of a bacterial population subjected to natural selection and mutation. In particular we will propose a theoretical model describing some particular features of E.coli, a bacterial species whose biological mechanisms are known with a quite good precision.

In a growing population, the individuals that are more adapted to the environment and to its possible changes are selected. The most suitable individuals must be able not only to duplicate fastly but also to react to eventual environmental changes, that is they must have a certain variability. Variability is provided by the processes of mutation and gene activation. In our model we will consider only the process of mutation, occuring during the evolution in a certain fixed environment.

Natural selection acts at all levels of the genetic information processing and biological organization: from DNA stability, replication and transcription to RNA translation efficiency to the correct functioning of the gene products. In principle all these contstraints could interact in a very complex way. Nevertheless it is worth trying to investigate the role of each element separately.

The fitness for a bacteria depends on a lot of factors: translation velocity, adaptability to environmental changes, antibiotic resistance, accuracy in duplicating. All these contstraints can be incompatible, for example having a lot of genes means low duplication, and low mutation rate means a lot of "right descendant", but also low adaptability. The problem in doing a theoretical model taking into account such effects, is to measure the fitness and its variation as a mutation occurs (that is the effect on the protein functionality by changing one codon).

Due to the degeneration of genetic code different codons, called synonymous, code for the same amino acid. The bias in codon usage is not random and codons preferentially used in highly espressed genes are translated by the most abundant isoacceptor tRNA. Codon bias is thought to be an effect of improvement in the efficiency of translation process under natural selection.

Synonymous mutations are those that are easier to treat, because do not affect the protein functionality. Nevertheless they act on the translation time since synonymous codons are related to different tRNAs. Such topic, that is the effect of synonymous mutations, is useful to confirm the scheme order efficiency, velocity against casuality and versatility.

We develop a model in which the genetic code of bacteria is represented by binary sequences, consisting of two synonymous codons, 0 and 1, that is DNA's triplets coding for the same amino acid. We are able to write a sort of *master equation* for the distribution of the different strains in the population. Since the Boltzmann distribution is a solution of such an equation, we have identified the competition between optimization and disorder balanced by mutations with the process of minimization of free energy in statistical systems. By means of this analogy, we can suggest an hypothesis and suppose that the system can be described by a Hamiltonian by using the spin formalism. In this way we obtain a system which is formally similar to a one-dimensional antiferromagnetic Ising system. Although the model can be described by means of statistical mechanics tools, the Hamiltonian will be in general more complex: we will try to see to what extent a simple Hamiltonian gives qualitatively correct results.

Due to the approximations made in the model, the direct comparison of the model with the available experimental data is only to some extent possible. Therefore in order to verify the theoretical predictions of the model it will be necessary to simulate the evolution of the system.

2 Biology

Before developing our model, let us give some biological definitions concerning cellular components and processes that will be used for developing the model. Although such an explanation cannot be exhaustive, nevertheless it will be useful for physicists who do not have any biological background.

2.1 Bacteria

Bacteria are unicellular organisms of microscopic dimensions (a few micron). Escherichia coli (E.coli) is a largely widespread bacterial species. An E.coli bacteria has rigid walls, is formed by a cytoplasmatic membrane containing genetic material immersed in cytoplasm. A bacteria reproduces by binary division: after having reached a sufficient size it duplicates all the cellular components and divides into two identical cells. The duplication time depends on the species and on the environmental conditions; it can varies from 30 minutes and a few hours. In a bacterial culture there are about 10^{12} cells.

2.2 Proteins

Proteins are one of the most important factors for the functions of the cell (substances exchange with environment, control and regulation of genic expression). Proteins are one-dimensional chains consisting of 20 different amino acids. Amino acids are organic acids containing a carboxyl $-COOH$ and an amino group $-NH_2$. The sequence of amino acid of a protein is coded by the genetic code of the cell. Amino acids are linked by a peptide bond binding the carboxyl group of an amino acid and the amino group of the next one. Protein's length goes from 40 up to 1000 amino acids.

2.3 Nucleic acids

Nucleic acids constitute genome, where genetic information is stored. In the cell there are two types of nucleic acids: *deoxyribonucleic acid* (DNA) and *ribonucleic acid* (RNA).

Nucleic acids are macromolecules (*polynucleotides*) consisting of 4 units called *nucleotides*. A nucleotide is characterized by an organic base: a *purine* (*adenine* or *guanine*) or a *pirimidine* (*citosine* or *timine* for DNA, for RNA *uracil* substitutes *timine*). Bases are abbreviated with A, G, C, U, T. These letters usually indicate also the corrisponding nucleotides. The sequence of nucleic acid coding for a protein is called *gene*.

DNA consists of two nucleotidic chains forming a double helix. Purine and pirimidines are bound by means of hydrogen bonds. Due to the chemical structure of nucleotides A and G on a chain are bound with T and C respectively on the other chain.

There are three different types of RNA in the cell:

- *Messenger* RNA (mRNA) contains the genetic information directly transcripted from DNA.

- *Transfer* RNA (tRNA) plays the role of adaptor, and leads the amino acid to the site where protein is synthetized according to the DNA information.

- *Ribosomal* RNA (rRNA) forms, together with ribosomal proteins, ribosomes, that are able to bind all the molecules involved in the proteic synthesis, to move along the mRNA chain and to build up the proteic chain. This process is called *translation*, because a proteic chain composed by amino acid is built, by reading a nucleic acid sequence.

2.4   The code

The genetic code is formed by bases triplet called *codons*. The total number of different codons is 64 (4^3). 61 of the 64 different codons code for an amino acid. The remaining three amino acids correspond to termination signals and indicate the end of the chain. Since the number of possible codons is larger than the number of amino acids, the code is degenerate, that is different codons (called *synonymous*) code for the same amino acid. The starting signal is AUG and corresponds to the amino acid Methionine.

2.5   Proteic syntesis

Code's translation, that is connection between codons on mRNA and related amino acids on protein does not take place directly. tRNA acts as adaptor: it is able to bind an amino acid on one side and to recognize the related mRNA codon on the other side where the conjugate triplet of the related codon (*anticodon*) is located. This way the tRNA brings the amino acid near the translation machinery (ribosome, mRNA and specific factors) so that the amino acid can bind the growing proteic chain.

In order to act as adaptor, tRNA must be activated by one of the 20 enzymes called aminoacyl-tRNA syntetase.

In bacteria there are 50 different types of tRNA. tRNAs molecules are generally composed by 70–80 bases. Due to phenomenon of *wobble*, that is non standard pairing of bases (different from A-T and G-C) in codon-anticodon coupling, each tRNA is able to bind more than one codon (but not necessarily all) between those corresponding to its related amino acid. Different tRNAs bind together with the common codon with different efficiencies. Ribosomes are complexes formed by RNA and more than 50 types of proteins.

The process of proteic syntesis can be described by distinguishing three different phases: start, elongation, termination.

Start

Starting signal for each proteic chain is AUG (corresponding to methionine). Ribosome binds to mRNA near to the starting signal. During this phase the complex $tRNA_{met}$, ribosome and mRNA is formed. $tRNA_{met}$ is located on a particular site on the ribosome called site P.

Elongation

If an activated tRNA corresponding to the second codon (after AUG) approaches to the complex, it can bind the ribosome in a site next to site P called site A. A peptide bond is built up between the two amino acids while the tRNA related to the second amino acid moves to the site P and tRNA$_{met}$ is released without its amino acid. For each translated codon the complex moves (*translocates*) of three bases. Translation process goes on until that the "translation machinery" meets a termination signal.

Useful data of tRNA cycle are those obtained from [1]. For a growing rate of 2 duplication/h, the cycle's duration is 0.69 s. 17% of this time (0.12 s) is spent for activation. An activated tRNA spends the most of its cycle time in searching for the right site A.

Codon usage influences the rate of translation, being the pairing between codons and diffusing tRNAs the rate limiting step compared with the attachment of the amino acid to the growing protein and the ribosome's translocation. In other words, the chain's elongation rate depends on the number of "right" tRNAs colliding with their related ribosomal site per unit of time.

The diffusion velocity of activated tRNAs can be considered instantaneous if compared with their cycle time. During the growing phase, more tRNAs are simultaneosly translated by more than one ribosome.

Termination

As soon as a ribosome reaches a termination signal the complete protein is released by the translation machinery.

2.6 Mutation

During bacterial growth, cell divides into two identical cells each of which receives a copy of the genome. Mistakes in DNA duplication are called *mutations*.

Genic mutations are particular mistakes involving one base only. Between mutations we will consider base substitutions not influencing protein's functionality, that is mutations transforming a codon in a synonymous one.

The *mutation rate* in bacteria is defined as mutation probability per time unit. It is obtained by dividing the number of mutations with the number of generations in which mutations occur. Mutation rate goes generally from 10^{-4} to 10^{-9}.

2.7 Mutation and selection

Mutation and selection take place during each step of bacteria's life, from proteic syntesis to protein functionality. Selection is the result of a complex interaction of many factors. Individuals are selected according to their capacity to survive in the environment or to better adapt to environmental changes. In our model we will investigate selection and mutation occuring during translation. Codon usage in bacteria is not random. For highly expressed genes it depends on tRNAs concentration in the cell, so that the codons are translated by the most abundant tRNAs. In this way proteic syntesis is optimized, since the wait of the translation machinery for the right activated tRNA is the rate limiting step of translation. Preferred codons used in genome between their synonymous ones are called *major codons.*

The so called *growth maximization strategy* says that in order to reach optimal conditions of synthesis, for gene sequences playing an important role during the phase of fast growth (as ribosomal genes), major codons are those corresponding to the most abundant tRNA. This way the cell obtains a strong selective advantage if compared with cells having a random distribution of codons in genome.

3 The model

3.1 General features

In order to study the system theoretically[3], let us make some hypothesis and approximations that can be to some extent biologically justified.

We will consider a highly expressed protein, so that we will be able to assume the translation time of the relative mRNA to be proportional to the cell duplication time.

We will disregard any spatial gradient in the concentration of tRNAs, that is we will suppose that "fast" diffusion of the molecules takes place in the cell.

Finally we will suppose that all available tRNAs are charged, that is we will suppose that the efficency of aminoacyl-tRNA synthetases is so high that the concentration of charged tRNA does not vary during translation.

Let us consider an mRNA chain of length l composed by two types of synonimous codons: 0 and 1. Let us suppose that the codons translation time is inversely proportional to the aboundance of the relative tRNAs. If a_0 and a_1 are the number of available tRNA respectively related to codons 0 and 1, the mean translation time for the whole chain is

$$\tau = \frac{j}{a_1} + \frac{l-j}{a_0} = rj + q,$$

where j is the number of codon of type 1, $r = (a_0 - a_1)/a_0 a_1$, $q = l/a_0$. Let us call M_j the mass of bacteria with the same number j of codons of type 1 which are present in $g_j = \binom{l}{j}$ different strains. Bacteria belonging to the same group j have the same duplication time, since they have the same magnetization.

The exponential growth of the mass M_j of the group j is

$$\frac{dM_j}{dt} = \nu_j M_j, \tag{1}$$

where $\nu_j = \tau_j^{-1}$ is the rate of duplication. We will consider only synonimous mutations occuring with mutation rate μ (the same for all codons). If we suppose a very low rate of mutation, we could think that only one mutation (from 1 to 0 or from 0 to 1) occours in one generation If we call μ/l the rate of mutation per codon, with $\mu \ll 1$, the average relative number of individuals of the group j undergoing to mutation per unit of time is $\mu \nu_j$. By introducing the average mass of a strain $m_j = M_j/g_j$ we obtain

$$\frac{dm_j}{dt} = (1 - \mu)\,\nu_j m_j + \frac{j}{l}\mu\nu_{j-1}m_{j-1} + \frac{l-j}{l}\mu\nu_{j+1}m_{j+1}, \tag{2}$$

where $0 \leq j \leq l$ and we have approximated $\nu_{-1} = \nu_{l+1} = 0$.

In a certain environment, a bacterial population grows as long as resources are available. More simply, we can consider a situation in which the population is fixed and nevertheless each bacteria grows exponentially. In such a condition we suppose that the selection acting as far as the velocity in duplicating is concerned is stronger, while in the *starving phase* the selection acts principally on the genic availability, since during such phase bacteria must be able to synthetiza a larger amount of different proteins. A natural way to describe mathematically this alternation is consider a normalized distribution and assuming that individuals favoured by natural selection are those which are able to duplicate faster.

By multiplying with g_j both members of (2) and summing up we obtain

$$\frac{dM}{dt} = \sum_{k=0}^{l} g_k \nu_k m_k = M \sum_{k=0}^{l} g_k \nu_k p_k = M \bar{\nu}$$

($M = \sum_{k=0}^{l} g_k m_k$), because mutations do not change the total mass of population. We are able to get the evolution equation for the probability distribution writing the coupled maps equation

$$\frac{dp_j}{dt} = \frac{d}{dt}\left(\frac{m_j}{M}\right) = \frac{1}{M}\frac{dm_j}{dt} - \frac{m_j}{M^2}\frac{dM}{dt},$$

so that

$$\frac{dp_j}{dt} = [(1 - \mu)\,\nu_j - \bar{\nu}]\,p_j + \frac{j}{l}\mu\nu_{j-1}p_{j-1} + \frac{l-j}{l}\mu\nu_{j+1}p_{j+1}, \tag{3}$$

where $\bar{\nu} = \sum_{k=0}^{l} g_k \nu_k p_k$.

Let us call the coefficient of p_j in (3) "fitness function" of the system, since it is an increasing function of ν_j, that is the capacity of the cell of duplicating fastly in order to survive better in its environment.

We could think at the bacterial population as a statistical mechanical system, in which the minimization of internal energy (corresponding to duplication time) is opposed by the tendency to the maximal entropy in order to reach the minimal free energy. The temperature (relating to mutation rate) acts as a balancing term.

By taking into account the contstraints for the normalization of the probability distribution $\sum_{k=0}^{l} g_k p_k$ in the maximization of "entropy"

$$S = -\sum_{k=0}^{l} g_k p_k \ln p_k$$

between bacteria having the same fitness ($S = -\sum_{k=0}^{l} g_k p_k \tau_k$) we obtain for the asimptotic state the Boltzmann distribution

$$p_j = C \exp\left(-\beta \tau_j\right), \tag{4}$$

where β can be considered as the inverse of an effective temperature, which in this case can be expressed by means of the other parameters

$$T = \frac{4\epsilon}{(1-\epsilon)} (ln(x))^{-1}, \tag{5}$$

where

$$\epsilon = a_1 - a_0,$$
$$x = \frac{(1-\mu)\epsilon + sqrt(1-\mu)^2 \epsilon^2 + \mu^2(1-\epsilon^2)}{\mu(1-\epsilon)}.$$

3.2   Simulations

In modelling biological systems, we are facing with the problem of discrepancy between their real structure and the corresponding model. In our case, complexity of bacteria prevents us doing a direct comparison with biological data. Therefore, in order to check the assumptions and verify the predictions of the model, we have simulated bacterial evolution.

In simulations, a population is composed by N bacteria, each of which is represented by a four elements-structure: an mRNA chain composed by codons 0 and 1; a tRNA pool of the two related species (a_0 and a_1); a ribosome translating the chain indicated by its position i ($i = 1, 2, ..., l$ where l is the length of the chain); a time index t indicating the time passed since translation's start. At the beginning of translation all mRNA chains are different and are randomly composed by 0 and 1. For each structure the same number of initial tRNA a_1^i and a_0^i is available. The ribosomes position is on the first codon of the relative chain. The system evolves for a time t_{max}.

At each temporal step the ribosome, located in the position i, translates the codon i, with a probability proportional to the number of the codon's related activated tRNAs. After that the ribosome has reached the l position, the corresponding bacteria duplicates, since we suppose that the mRNA chain codes for a highly expressed gene, so that we can assume that translation time and duplication time are proportional. A bacteria is randomly chosen in the population and each element of its structure is substituted by the elements of the duplicating one. This way the number of bacteria in the population is constant. After duplicating, the copied bacteria mutates: each codon of the chain changes $(0 \to 1$ or $1 \to 0)$ with probability μ. Finally a bacteria restart to translate.

The simulated system is more general that the analytical one. In the analytical model reactivation time is constant and equals the traslocation time, that is a codon interacts always with his previous neighbour. In the simulation mechanism, reactivation time varies from site to site; this way the number of interacting codons is not constant. In order to calculate the interesting quantities we suppose that the simulated system is ergodic. This assumption is reasonable since that all the regions of the phase space can be explored (all the configuration are a priori allowed). Growth optimization is generally related to the problem of minimization of a function (energy in equilibrium statistical systems) under certain contstraints and with chaos (temperature). During the evolution, the system approaches to a minimal energy configuration. For vanishing temperature we can imagine the search of the optimization as the approaching, step by step to a state of smaller energy. If the energy fuction has more than one minimum the system can remain trapped in a local minimum, separated from the absolute one by one or more barriers. If temperature is larger than zero the system becomes able to cross such barriers and to explore a larger region of the phase space. If μ is sufficiently large, our system visits easly all the phase space but once he has reached the minimal energy configuration, it does not necessarily remains there. If initially temperature is mantained high, the system gains access to a large region of the space without being trapped. If temperature is gradually decreased it becomes able to distinguish between high and low energy configurations since the ratio energy/temperature changes in a more sensitive way by varying configuration. By further decreasing temperature we force the system to remain in the deepest valley of energy and to reach more easly the absolute minimum. This method is called *annealing*. In analogy with this method we can simulate a process in which temperature is gradually decreased in time in order to favourite the reaching of the minimum (*simulated annealing*). In our system every configuration has a related translation time corresponding to the function to be minimized: in fact we are interested in duplication rate optimization for given parameters. Mutations vary configurations in a discrete way. Mutations transforming chains in more optimized codes are mantained in the population since a lower duplication time allows a larger number of duplication.

3.3 Statistical mechanics

We could think at the bacterial population as a statistical mechanical system tending to the minimization of the free energy F:

$$F = U - TS \tag{6}$$

Since the aim of a bacteria in order to get a selective advantage is to reach a fast growing rate, the duplication time could be analogous to internal energy. The temperature in this scenario is a control parameter connected with the mutation rate μ and with the *importance* of the protein, (that is the quantity of protein that must be synthetized by bacteria to duplicate).

Mutations make the codons distribution omogeneous, increase the disorder of the chain and, having the opposite effect to natural selection, favour the increasing of entropy. Therefore temperature must increase in dependence on μ. On the other side, for very important proteins should be very necessary optimize the duplication time: the more the protein is important, the smaller must be the effect of T. This way in the minimization process of F, balanced by T, for the same mutation rate, the translation time for important proteins is more "optimized".

Mutations are obviously a necessary factor for the reaching of an optimized distribution.

By carrying on the analogy duplication time-energy, we are able to identify our model with a 1D Ising model without interaction. In fact, by expressing the duplication time by means of a spin formalism (by substituting $s_i = (\sigma_i + 1)/2$) we obtain

$$\mathcal{H} = \frac{r}{2} \sum_{i=1}^{l} \sigma_i + q + \frac{r}{2}$$

Table 1: The abundance of tRNA for the four possible couples of codons

s_{i-1}	s_i	$(\tau(s_{i-1}, s_i))^{-1}$
0	0	$a_0 - 1$
0	1	a_1
1	0	a_0
1	1	$a_1 - 1$

The form of the Hamiltonian, an Ising Hamiltonian in external field without interaction, is due to the fact that we have a linear dependency of replication time on j, and the Boltzmann distribution (4) is a solution of the stationary equation.

3.4 Model with one ribosome, one mRNA

Starting from the analogy replication time-energy, we can introduce further terms in the hamiltonian simply by taking into account more element which we neglected in the first approximation in expressing translation time. Let us now consider as an example the case in which the tRNAs' reactivation time is not negligible. Let us suppose that a codon has been just translated and the translation machinery has just translocated in order to translate the next codon, identical to the previous one. If the reactivation time τ_r (identical for the two tRNAs 0 and 1) is larger than the codon translation time for the second codon a smaller number of activated tRNA (one charged tRNA less than before) will be available. In this case the translation time depends not only on the codon itself but also on its previous neighbour. Therefore for a codon s_i there are 4 (2^2) different possible cases, as shown in table 1.

In other words the translation time of a codon is proportional to the inverse of the initial number of tRNAs if the codon is different from the previous one, otherwise it is proportional to the inverse of the same number minus one since the related tRNA just used is still discharged. We can write τ_i as

$$\tau_i = \theta(s_{i-1}, s_i) =$$
$$\theta(0,0)(1 - s_{i-1})(1 - s_i) + \theta(0,1)(1 - s_{i-1})s_i +$$
$$\theta(1,0)s_{i-1}(1 - s_i) + \theta(1,1)s_{i-1}s_i.$$

Developing the eq. (7), substituting $s_i = (\sigma_i + 1)/2$ and summing over i we obtain [2]

$$\mathcal{H} = \mathcal{H}_0 - \frac{H}{l}\sum_{i=1}^{l}\sigma_i - \frac{J}{l}\sum_{i=0}^{l}\sigma_i\sigma_{i-1}, \tag{7}$$

where

$$H = \frac{1}{2}\left(\frac{1}{a_0 - 1} - \frac{1}{a_1 - 1}\right),$$

$$J = -\frac{1}{4}\left(\frac{1}{a_0(a_0 - 1)} + \frac{1}{a_1(a_1 - 1)}\right).$$

$\mathcal{H}$ is a constant term, H is the external fields and J is the coupling term. Since J is negative ($a_0 > 1$, $a_1 > 1$), the system is antiferromagnetic. Further we will investigate only the case $a_1 > a_0$, since the system is symmetric for changes $0 \rightleftharpoons 1$, so that we can consider the case $H > 0$. In order to investigate this antiferromagnetic system, let us introduce in the Hamiltonian a "staggered external fields" H_s

$$\mathcal{H} = \mathcal{H}_0 - \frac{H}{l}\sum_{i=1}^{l}\frac{\sigma_i + \sigma_{i-1}}{2} - \frac{H_s}{l}\sum_{i=1}^{l}\frac{|\sigma_i - \sigma_{i-1}|}{2} - \frac{J}{l}\sum_{i=0}^{l}\sigma_i\sigma_{i-1}. \quad (8)$$

The staggered magnetization is defined as

$$m_s = \sum_{i=1}^{l}\frac{|\sigma_i - \sigma_{i-1}|}{2l} = -\frac{\partial f}{\partial H_s}\bigg|_{H_s=0}, \quad (9)$$

where f is the free energy per spin. The staggered magnetization vanishes for a chain in a ordered ferromagnetic configuration ($m = \lim_{l\to\infty} m^{(l)} = 1$), is $1/2$ for a random configuration and ($m = 0$) and 1 in the case of a ferromagnetic configuration ($m = 0$). By means of the partition function we can calculate the magnetization and the staggered magnetization

$$m = \frac{\partial f}{\partial H_s}\bigg|_{\substack{H_s=0 \\ H=0}} = \left[\frac{sinh(\beta H)}{\sqrt{cosh^2(\beta H) - 2exp(-2\beta J)sinh(2\beta J)}}\right]_{\substack{H_s=0 \\ H=0}} = 0;$$

$$(10)$$

$$m_s = -\frac{\partial f}{\partial H_s}\bigg|_{H_s=0} = \frac{exp(-4\beta J)/\sqrt{cosh^2(\beta H) - 1 + exp(-4\beta J)}}{cosh(\beta H) + \sqrt{cosh^2(\beta H) - 1 + exp(-4\beta J)}}. \quad (11)$$

It is useful to study the behavior of the system by introducing the two variables S and D ($S = a_1 + a_0$; $D = a_1 - a_0$). In fig... the staggered magnetization as a function of S for different values of D is shown. The dotted line divides the physical region from the non physical one, (because of the contstraints $a_0 > 1$, $a_1 > 1$ it must be that $0 \leq D \leq S - 2$, so that the minimal value of S increases for growing D.

For large S, the coupling effect between the spins disappeard: the tRNA's *finite size effect* vanishes in this case and the selective advantage of the ordered chains is weak (either ferro or antiferromagnetic.

For small S, there is a strong competition between the antiferromagnetic interaction and external field favouring the presence of codons of type 1. If D is small antiferromagnetic configurations ($m_s > 1/2$) are the most probable. In this conditions a bacteria having the majority of codons of the type 1 does not obtain advantage, because the amount of the two tRNAs pools are similar. By increasing D the ferromagnetic m_s approaches to $1/2$ and for $D \approx 1.41$ the ferromagnetic configurations get a greater and greater advantage than the ferromagnetic ($m_s < 1/2$). In this case, if the tRNA pools are sufficiently small, the external field H is able to move the system in the ferromagnetic region. The greater is D, the longer this ferromagnetic effect is advantaged by increasing S.

Considering different reactivation times or investigating the two-ribosomes translation (by means of the same method used to obtain the (7)), we could obtain more complex Hamiltonians, with more-neighbours and long range interactions.

3.5 Comparison with biological data

From the simulations, we can get a "numerical" value for the quantity: magnetization, staggered magnetization, average duplication time together with related standard deviations. We can get also the population distribution at the end of the evolution process. In the case of neglected reactivation time, the population distribution allows us to get a numerical value of T, since this is the only situation in which a theoretical expression for the temperature is available.

References

1. M. Gouy and C. Grantham *Febs. Letters* **115**, 151-155 (1980).
2. F. Bagnoli, G. Guasti and P. Liò, *Translation Optimization in Bacteria: Statistical Models*, in *Non Linear Excitations in Biomolecules*, M. Peyrard (ed.), Springer-Verlag Berlin (1995) p. 405
3. F. Bagnoli and P. Liò *J. Theor. Biol.* **173**, 271-281 (1995).

AUTOMATA NETWORK MODELS IN ECOLOGY AND EPIDEMIOLOGY

NINO BOCCARA

Department of Physics, University of Illinois, Chicago, USA
and
DRECAM/SPEC CE Saclay, 91191 Gif-sur-Yvette Cedex, France
e-mail: boccara@uic.edu
home page: http://tigger.uic.edu/~boccara/nbhome.html

1 Introduction

The first task that faces the theoretician who wants to interpret the time evolution of a complex system is the construction of a model. In the actual system many features are likely to be important. Not all of them, however, should be included in the model. Only the few relevant features which are thought to play an essential role in the interpretation of the observed phenomena should be retained. Such simplified descriptions should not be criticized on the basis of their omissions and oversimplifications. The investigation of a simple model is often very helpful in developing the intuition necessary for the understanding of the behavior of complex real systems. In many-body physics, for instance, models such as the van der Waals model of a fluid, the mass and spring model of lattice vibrations, and the Ising model of cooperative phenomena, to mention just a few, have played a major role. A simple model, if it captures the key elements of a complex system, may educe highly relevant questions.

This lecture is devoted to the investigation of automata network models of interacting populations such as competing species in ecology or susceptibles and infectives in epidemiology.

1.1 Population oscillations in ecology

Most models in population dynamics are formulated in terms of differential equations, [a] the classical example being the predator-prey model proposed in the 1920's, independently, by Lotka[2] and Volterra.[3][b]

---

[a]For a rich and fascinating variety of models refer to Murray (1989) p. 659.[1]

[b]Vito Volterra (1860–1940) was stimulated to study this problem by his future son-in-law, Umberto D'Ancona (1954), who, analyzing market statistics of the Adriatic fisheries, found that, during the First World War, certain predaceous species increased when fishing was severely limited. A year before, Alfred James Lotka (1880–1949) had come up with an almost identical solution to the predator-prey problem. His method was very general, but,

This model assumes that, in the absence of predators, the prey population, denoted by H for "herbivore", grows exponentially, whereas, in the absence of preys, predators starve to death and their population, denoted by P, declines exponentially. As a result of the interaction between the two species, H decreases and P increases at a rate proportional to the frequency of predator-prey encounters. We have then

$$\dot{H} = bH - sHP$$

$$\dot{P} = -dP + esHP,$$

where b is the birth rate of the prey, d the death rate of the predator, s the searching efficiency of the predator and e the efficiency with which extra food is turned into extra predators. [c] These equations may be reduced to a dimensionless form. If we put

$$h = \frac{Hes}{d}, \quad p = \frac{Ps}{b} \quad \tau = \sqrt{bd}\,t, \quad \rho = \sqrt{\frac{b}{d}},$$

we obtain

$$\frac{\mathrm{d}h}{\mathrm{d}\tau} = \rho h\,(1 - p)$$

$$\frac{\mathrm{d}p}{\mathrm{d}\tau} = -\frac{1}{\rho}p\,(1 - h).$$

These equations contain only one parameter.

The Lotka–Volterra model has two equilibrium states (0,0) and (1,1). (0,0) is unstable and (1,1) is stable but not asymptotically stable. Except the coordinate axes and the equilibrium point (0,0), all the trajectories are closed orbits oriented counterclockwise.

Since the trajectories in the prey-predator phase space are closed orbits, the populations of the two species are periodic functions of time. This result is encouraging because it might point toward a simple relevant mechanism for predator-prey cycles. There is an abundant literature on cyclic variations of animal populations (see, in particular, Finerty (1980) [7]). They were first observed in the records of fur-trading companies. The classic example is the records of furs received by the Hudson Bay Company from 1821 to 1934. They

---

probably because of that, his book—reprinted as *Elements of Mathematical Biology* [4]—did not receive the attention it deserved. On the relations between Lotka and Volterra, and how ecologists in the 1920s perceived mathematical modeling, consult Kingsland. [5]

[c]Chemists will note the similarity of these equations with the rate equations of chemical kinetics. For a treatment of chemical kinetics from the point of view of dynamical systems theory, consult Gavalas. [6]

show that the numbers of snowshoe hares ^{d} (*Lepus americanus*) and Canadian lynx (*Lynx canadensis*) trapped for the company vary periodically, the period being of about 10 years. The hare feeds on a variety of herbs, shrubs and other vegetable matter. The lynx is essentially single-prey oriented, and although it consumes other small animals if starving, it cannot live successfully without the snowshoe hare. This dependence is reflected in the variation of lynx numbers which closely follow the cyclic peaks of abundance of the hare usually lagging a year behind. The hare density may vary from one hare per square mile of woods to 1000 or even 10000 per square mile. ^{e} In this particular case, however, the understanding of the coupled periodic variations of predator and prey populations seems to require a more elaborate model. The two species are actually parts of a multi-species system. In the boreal forests of North America, the snowshoe hare is the dominant herbivore, and the hare-plant interaction is probably the essential mechanism responsible for the observed cycles. When the hare density is not too high, moderate browsing remove the annual growth and has a pruning effect. But at high hare-density browsing may reduce all new growth for several years and, consequently, lower the carrying capacity for hare. The shortage in food supply causes a marked drop in the number of hares. It has also been suggested that when hares are numerous, the plants on which they feed respond to heavy grazing by producing shoots with high levels of toxins. [9] If this interpretation is correct, the hare cycles would be the result of the herbivore-forage interaction (in this case, hares are "preying" on vegetation), and the lynx, because they depend almost exclusively upon the snowshoe hares, track the hare cycles.

Since, in nature, the environment is continually changing, in phase space, the point representing the state of the system will continually jump from one orbit to another. From an ecological viewpoint, an adequate model should not yield an infinity of neutrally stable cycles but one *stable limit cycle*. That is, in the (h, p) phase space, there should exist a closed trajectory C such that any trajectory in the neighborhood of C should, as time increases, become closer and closer to C.

Furthermore, the Lotka–Volterra model assumes that, in the absence of predators, the prey population grows exponentially. This Malthusian growth ^{f}

---

^{d}Also called varying hares, have large heavily furred hind feet and a coat that is brown in summer and white in winter.

^{e}Many interesting facts concerning northern mammals may be found in Seton (1909). [8]

^{f}After Thomas Robert Malthus (1766–1834) who, in his most influential book, [10] stated that, because a population grows much faster than its means of subsistence—the first increasing geometrically whereas the second increase only arithmetically—"vice and misery" will operate to restrain population growth. To avoid these disastrous results, many demographers, in the nineteenth century, were lead to advocate birth control. More details on Malthus and is not realistic. Hence, if we assume that, in the absence of predation, the growth of the prey population follows the logistic model, we have

$$\dot{H} = bH \left(1 - \frac{H}{K}\right) - sHP$$

$$\dot{P} = -dP + esHP,$$

where K is the carrying capacity of the prey. If, to the dimensionless variables defined above, we add the scaled carrying capacity

$$k = \frac{Kes}{d},$$

the equations may be written

$$\frac{dh}{d\tau} = \rho h \left(1 - \frac{h}{k} - p\right)$$

$$\frac{dp}{d\tau} = -\frac{1}{\rho}p\left(1 - h\right).$$

For large k, this model is a small perturbation of the Lotka–Volterra model. The equilibrium points are $(0,0)$, $(k,0)$ and $(1, 1 - 1/k)$. Note that the last equilibrium point exists if, and only if, $k > 1$, that is, if the carrying capacity of the prey is high enough to support the predator. $(0,0)$ and $(k,0)$ are unstable while $(1, 1 - 1/k)$ is stable. A finite carrying capacity for the prey transformed the neutrally stable equilibrium point of the Lotka–Volterra model into an asymptotically stable equilibrium point.

A small perturbation—corresponding to the existence of a finite carrying capacity for the the prey—has qualitatively changed the phase portrait of the Lotka–Volterra model. A model whose qualitative properties do not change significantly when it is subjected to small perturbations is said to be *structurally stable*. Since a model is not a precise description of a system, qualitative predictions should not be altered by slight modifications. Satisfactory models should be structurally stable.

If we limit our discussion of predation to two-species systems assuming, as we did sofar, that

(i) the time t is a continuous variable,

(ii) there is no time lag in the responses of either populations to changes,

(iii) population densities are not space-dependent,

---

his impact, may be found in Hutchinson.[11]

a somewhat realistic model, formulated in terms of ordinary differential equations, should, at least, take into account the following relevant features: [12,13]

1. *Intraspecific competition*, that is, competition between individuals belonging to the same species.

2. *Predator's functional response*, that is, the relation between the predator's consumption rate and prey density.

3. *Predator's numerical response*, that is the efficiency with which extra food is transformed into extra predators.

Essential resources being limited in general, intraspecific competition reduces the growth rate, which eventually goes to zero. The simplest way to take this feature into account is to introduce in the model carrying capacities for both the prey and the predator.

A predator has to devote a certain time to search, catch and consume its prey. If the prey density increases, searching becomes easier, but consuming a prey takes the same amount of time. The functional response is, therefore, an increasing function of the prey density—obviously equal to zero at zero prey density—approaching a finite limit as high densities. In the Lotka—Volterra model the functional response, represented by the term sH, is not bounded. The behavior of the functional response at low prey density depends upon the predator. [14,15]

In the Lotka–Volterra model, the predator's numerical response is a linear function of the prey density. As for the functional response, it can be argued that there should exist a saturation effect, that is, the predator's birth rate should tend to a finite limit at high prey densities.

To conclude this discussion of population oscillations, let us briefly describe an interesting predator-prey model due to Tanner. [16]

The equation for the prey is

$$\dot{H} = r_H H \left(1 - \frac{H}{K} \right) - \frac{aPH}{b + H}.$$

The growth of the prey in the absence of predators is modeled by the logistic equation. r_H is the intrinsic growth rate and K is the carrying capacity. The predator's functional response behaves linearly at low prey densities, and tends to a at high prey densities.

The equation for the predator is

$$\dot{P} = r_P P \left(1 - \frac{P}{cH} \right).$$

It has the logistic form with a carrying capacity for the predator proportional to the prey density. This model contains 6 parameters. Since there is only one nontrivial equilibrium point (H^*, P^*), [9] if we put

$$h = \frac{H}{H^*}, \quad p = \frac{P}{P^*}, \quad \rho = \sqrt{\frac{r_H}{r_P}}, \quad k = \frac{K}{H^*}, \quad \beta = \frac{b}{H^*}, \quad \tau = \sqrt{r_H r_P}\, t,$$

The equations may be written

$$\frac{dh}{d\tau} = \rho h \left(1 - \frac{h}{k}\right) - \frac{\alpha p h}{\beta + h}$$

$$\frac{dp}{d\tau} = \frac{1}{\rho} p \left(1 - \frac{p}{h}\right).$$

These equations contain three independent parameters: ρ, k and β, the extra parameter α being given in terms of these by

$$\alpha = \rho \left(1 - \frac{1}{k}\right)(\beta + 1).$$

The equilibrium points are $(0,0)$, $(k,0)$ and $(1,1)$. $(0,0)$ and $(k,0)$ are unstable. $(1,1)$ is asymptotically stable if, and only if,

$$\rho^2(k - \beta - 2) < k(\beta + 1).$$

The nontrivial equilibrium point is always asymptotically stable if $k - \beta - 2$ is negative. If $k - \beta - 2$ is positive, the above condition may be written

$$\rho^2 < \frac{k(\beta + 1)}{k - \beta - 2}$$

which shows that, for fixed values of k and β, there exists a threshold

$$\rho_c = \sqrt{\frac{k(\beta + 1)}{k - \beta - 2}}.$$

For $\rho < \rho_c$, $(1,1)$ is asymptotically stable; for $\rho > \rho_c$, it is unstable, and there exists a stable limit cycle, that is, the populations of the two species are periodic functions of time.

---

[9] (H^*, P^*) is the unique solution of the system

$$r_H \left(1 - \frac{H}{K}\right) = \frac{aP}{b + H},$$

$$P = cH.$$

Note that $K > H^*$.

1.2 Epidemics models

The different models to be discussed here are extensions of the so-called "general epidemic model". [17] In this model, infection spreads by contact from infectives to susceptibles, and infectives are removed from circulation by death or isolation. The first and simplest model was proposed by Kermack and McKendrick. [18] A nice discussion of this model can be found in Waltman. [19] Kermack and McKendrick assumed that infection and removal were governed by the following rules:

(i) The rate of change in the susceptible population is proportional to the number of contacts between susceptibles and infectives, where the number of contacts is taken to be proportional to the product of the number of susceptibles S by the number of infectives I.

(ii) Infectives are removed at a rate proportional to their number I.

(iii) The total number of individuals $S + I + R$, where R is the number of removed infectives, is constant, that is, the model ignores births, deaths by other causes, immigration, emigration, etc.

If S, I and R are supposed to be real positive functions of time t, (i), (ii) and (iii) yield

$$\frac{dS}{dt} = -iSI$$

$$\frac{dI}{dt} = iSI - rI$$

$$\frac{dR}{dt} = rI,$$

where i and r are positive constants representing, respectively, the infection rate and the removal rate. From the first equation, it is clear that S is a nonincreasing function, whereas the second equation implies that $I(t)$ increases with t if $S(t) < r/i$ and decreases otherwise. Therefore, if, at $t = 0$, the initial number of susceptibles $S(0)$ is less than i/r, since $S(t) \leq S(0)$, the infection dies out, that is, no epidemic occurs. If, on the contrary, $S(0)$ is greater than the critical value i/r, the epidemic occurs, that is, the number of infectives first increases and then decreases when $S(t)$ becomes less than i/r. [h] This "threshold phenomenon" shows that an epidemic can occur if, and only if, the initial number of susceptibles is greater than a threshold value.

The Kermack-McKendrick model assumes a homogeneous mixing of the population, that is, it neglects the local character of the infection process. The

---

[h]That is, *by definition*, an epidemic occurs if dI/dt is positive at $t = 0$.

model also neglects the motion of the individuals, which is a factor that clearly affects the spread of the disease.

To take into account the motion of the individuals, it is usually assumed that they disperse randomly. This hypothesis amounts to incorporating diffusion terms in the equations. Models of this type help understanding the spatial spread of epidemics. Consider, for instance, rabies epidemic among foxes. Rabies is a viral infection of the nervous central system. It is transmitted by contact and is invariably fatal. If the virus enters the limbic system, that is, the part of the brain thought to control behavior, the fox loses its sense of territory and wanders in a more or less random way. To discuss the spatial spread of rabies among foxes in Europe, Källén et al. [20] added a diffusion term in rate equation of the infectives in the Kermack-McKendrick model in order to take into account the random dispersion of the rabid foxes. [i] We have then

$$\frac{\mathrm{d}S}{\mathrm{d}t} = -iSI$$

$$\frac{\mathrm{d}I}{\mathrm{d}t} = iSI - rI + D\frac{\partial^2 I}{\partial x^2}$$

$$\frac{\mathrm{d}R}{\mathrm{d}t} = rI,$$

where D is the diffusion coefficient of the infected foxes. This system of equations admit travelling wavefront solutions of the form $S(x - ct)$, $I(x - ct)$ and $R(x - ct)$, where c is the speed propagation of the epidemic wave. For the epidemic to occur, the average initial susceptible population density, i.e., ahead of the epidemic wave, has to be greater than the threshold value r/i; and, in this case, it is found that c behaves as $D^{1/2}$. To explain the observed fluctuations in the susceptible fox population density after the passage of the wavefront, Murray et al. [21] have considered a less simple model taking into account fox reproduction and the existence of a rather long incubation period (12 to 150 days).

Although the models presented sofar have unquestionably contributed to our understanding of the spread of an infectious disease (e.g., Murray's model allows for quantitative comparison with known data), the short-range character of the infection process is not correctly taken into account. This will be manifest when we discuss systems that exhibit bifurcations. In phase transition theory, for instance, it is well-known that in the vicinity of a bifurcation point—i.e., a second-order transition point—certain physical quantities have a singular behavior. [j] It is only above a certain spatial dimensionality—known

---

[i] See also Murray. [1]
[j] See, e.g., Boccara (1976) pp 155–189. [22]

as the upper critical dimensionality—that the behavior of the system is correctly described by a partial differential equation. For instance, the spatial fluctuations of the order parameter close to a second-order transition point are correctly described by the time-independent Landau–Ginzburg equation above 4 dimensions. [k]

One way to take correctly into account the short-range character of the infection process is to discretize space, and to represent the spread of an epidemic as the growth of a random cluster on a lattice. A kinetic model of cluster growth may be defined as follows[23,24] Denote, as usual, by Z^2 the two-dimensional square lattice. At a time t a site of Z^2 is either vacant (healthy), occupied (infected) or immune. An immune site is one which has been occupied in the past. At time $t + 1$ a vacant site becomes occupied with a probability p if, at least, one of its neighbor is occupied at time t. An occupied site at time t becomes immune at time $t + 1$. However, immunisation is not perfect and an immune site may become reoccupied with probability $p - q$ if, once again, one of its neighbors is occupied. More generally, one might assume that the probability that a site becomes occupied depends on the number of occupied neighbors. If $p = q$, any bond can be tried only once, since at a second try one of the neighboring sites is completely immune and no infection can pass. A similar model has been studied by McKay and Jan[25] to discuss forest fires. Vacant, occupied and immune sites correspond, respectively, to sites occupied by unburnt, burning and burnt trees. It is found that there is a critical probability p_c below which only a finite number of sites are immune. In the vicinity of p_c the system exhibits a second-order phase transition characterized by a set of critical exponents. The upper critical dimensionality is equal to 6, and Cardy[26] has calculated the critical exponents to first order in $\epsilon = 6 - d$. Cardy and Grassberger[24] have shown that these models are in the same universality class—i.e., have the same critical exponents—as percolation cluster growth models. The relationship of the general epidemic model to the percolation process has been first noticed by Mollison.[27]

The general epidemic model on a lattice may be viewed as a discrete dynamical system, in space and time. More precisely, it may be defined as a probabilistic automata network. In simple words, an automata network[28] consists of a graph where each site takes states in a finite set. The state of a site changes in time according to a rule which takes into account only the states of the neighboring sites in the graph. This is the point of view which will be adopted in these lectures.

To conclude this introduction, it is probably worthwhile to give a slightly more general definition of the spatial general epidemic model since, after the

---

[k]Ibid.[22] pp 227–274.

review paper of Mollison[27], several papers have appeared in the mathematical literature on this topic.[l] Let V be a set of sites (usually $V = \mathbb{Z}^d$). At any time $t \geq 0$ each site is either empty or has a healthy or an infected individual. The number of sites with infected individuals is initially finite. An infected individual emits germs in a Poisson process until he is removed after a random lifetime. Each germs goes independently to another site chosen according to a probability distribution attached to the parent site. If a germ meets an infected individual or goes to an empty site, nothing happens. After an individual has been removed his site remains empty for ever. The infectives have all the same emission rate and identical lifetime distribution.

All these different versions of the spatial general epidemic model still neglect the motion of the individuals. The influence of this factor on the spread of the epidemic is one of the main concerns of this lecture. Various models will be discussed. All of them are site-exchange cellular automata, that is, automata networks whose local rule consists of two subrules. The first one, applied synchronously, models the interaction process between the individuals. It is a probabilistic cellular-automaton rule. The second subrule, applied sequentially, models the motion of the individuals. It is a site-exchange rule. Such models may also be viewed as interacting particle systems. The mathematically-oriented interested reader should refer to Liggett (1985).[32]

2   Site-exchange Cellular Automata

Site-exchange cellular automata (CAs) are automata networks whose rule consists of two subrules. The first one is a standard synchronous CA rule, whereas the second is a sequential site-exchange rule. This last rule, characterized by a parameter m, is defined as follows. A site, whose value is one, is selected at random and swapped with another site value (either zero or one) also selected at random. The second site is either a neighbor of the first one (local site-exchange) or any site of the lattice (nonlocal site-exchange). This operation is repeated $mc_f(m,t)N$ times, where N is the total number of sites, $c_f(m,t)$ the density of nonzero sites at time t, and f is the CA rule. The parameter m is called the *degree of mixing*. It is important to note that this mixing process, which will be used to model either short- or long- range moves of interacting individuals, does not change the value of the density $c_f(m,t)$.

[l]See, e.g., Kuulasmaa (1982)[29], Kuulasmaa and Zachary (1984)[30] and Cox and Durrett (1988).[31]

2.1   The site-exchange rule as a mixing process

While short-range moves on a finite-dimensionnal lattice are clearly diffusive moves, long-range moves may be viewed as diffusive moves on an infinite-dimensional lattice. To help clarify the mixing process that results from the motion of the individuals, it is worthwhile to characterize the mixing process as a function of m and the lattice dimensionality d.

Consider a random initial configuration $C(0)$ of random walkers with density c on a d-dimensional torus Z_L^d. Select sequentially mcN walkers at random and move them to a neighboring site also selected at random if, and only if, the randomly selected neighboring site is vacant. *At random* means that all possible choices are equally probable. $N = L^d$ is the total number of sites of the torus and m, which is the average number of tentative moves per random walker, is a nonnegative real number. Let $C(mcN)$ denote the resulting configuration.

To characterize the mixing process, we consider the Hamming distance

$$d_H(C(0), C(mcN)) = \frac{1}{N} \sum_{j=1}^{N} (n(0,j) - n(mcN, j))^2,$$

where $n(0,j)$ and $n(mcN, j)$ are, respectively, the occupation numbers of site j in $C(0)$ and $C(mcN)$, as a function of the density c , the parameter m and the space dimensionality d.

If m is very large, $C(0)$ and $C(mcN)$ are decorrelated and the Hamming distance, which is the average value over space of

$$(n(0,j) - n(mcN, j))^2 = n(0,j) + n(mcN, j) - 2n(0,j)n(mcN, j)$$

is equal to $2c(1-c)$.

Since, in simulations, initial configurations are random, only averages over all random walks and all initial configurations with a density c of random walkers are meaningful quantities.

It can be proved (Boccara *et al.* (1994),[33] appendix) that, when N tends to ∞, the average reduced Hamming distance

$$\delta = \frac{d_H(C(0), C(mcN))}{2c(1-c)},$$

depends on m and d but not on c. For small m, δ behaves as m for all vallues of d whereasfor large m, $\delta - 1$ approaches 0 as $1/\sqrt{m^d}$.

2.2   Deterministic Site-Exchange Cellular Automata

If $m = \infty$, the correlations created by the application of the CA rule f are completely destroyed, and the value of the stationary density of nonzero sites $c_f(\infty, \infty)$ is then correctly predicted by a mean-field-type approximation in which it is assumed that the probability, at time t, for a site value to be equal to one is $c_f(m, t)$. This approximation is incorrect when m is not sufficiently large.

Deterministic cellular automaton evolution rules are defined as follows. The state at time t of a one-dimensional cellular automaton being represented by a function $S_t : i \mapsto s(t, i)$ from the set of integers Z to the set of states $\{0, 1\}$ the equation

$$s(t+1, i) = f\big(s(t, i-r), \ldots, s(t, i-1), s(t, i), s(t, i+1), \ldots, s(t, i+r)\big),$$

where f, which is a map from $\{0, 1\}^{2r+1}$ to $\{0, 1\}$, is the rule that determines the evolution of the range-r cellular automaton. $\{0, 1\}^{\mathsf{Z}}$ is the state space. An element of the state space is also called a configuration.

A study of the asymptotic behaviors for small and large degrees of mixing of three class-3 site-exchange cellular automata, namely radius-1 Rules 18, 54 and 22 (for rule code numbers see Wolfram 1983 and 1986[34,35]) shows[36] that for a small degree of mixing, the behavior depends upon whether the evolution of the deterministic cellular automaton may be viewed as annihilating interacting particlelike structures or not. [m] This result is pratically the same whether the exchange of site values is local or nonlocal. For a large degree of mixing, the asymptotic behavior for local and nonlocal site exchange is, on the contrary, completely different.

Another result, which clearly shows how the mean-field behavior is approached as the degree of mixing m is increased, is worth mentioning.[38] Consider a one-dimensional class-3 totalistic CA of radius r such whose evolution rule is given by

$$s(i, t+1) = \begin{cases} 1, & \text{if } S_{\min} \leq \sum_{j=-r}^{j=r} s(i+j, t) \leq S_{\max} \\ 0, & \text{otherwise}, \end{cases}$$

where r, $S_{\min}$ and $S_{\max}$ are chosen in such a way that the mean-fied map for the density of active sites is chaotic while the numerical simulation shows that the stationary density of active sites is a fixed point. The numerical study of the stationary density of active sites $c(m, \infty)$ as a function of m for long-range

---

[m] For more details on the evolution of particlelike structures, refer to Boccara *et al.* (1991).[3]

moves shows that $c(m, \infty)$ exhibits a sequence period-doubling bifurcations and behaves chaotically whem m is larger that a certain threshold.

One final remark. Note that in this section we considered deterministic CA rules, but taking into accont the site-exchange process, the resulting automaton is probabilistic.

2.3 Probabilistic Site-Exchange Cellular Automata

The evolution of the probabilistic cellular automaton determined by the following rule

$$
s(t+1, i) = \begin{cases} X & \text{if } s(t, i-1) = 0, s(t, i) = 0, s(t, i+1) = 1, \\ X & \text{if } s(t, i-1) = 1, s(t, i) = 0, s(t, i+1) = 0, \\ 0 & \text{otherwise,} \end{cases}
$$

where X is a Bernoulli random variable defined on $\{0, 1\}$ such that

$$
\Pr(X = 0) = 1 - p \quad \text{and} \quad \Pr(X = 1) = p,
$$

has been studied.[33] For $p = 1$, the above rule coincides with Rule 18.

For a fixed value of the degree of mixing m, the system exhibits, at $p = p_c(m)$, a transcritical bifurcation. The behavior of the stationary density of nonzero sites $c(m, p, \infty)$ in the neighborhood of the bifurcation point may be characterized by a critical exponent $\beta(m)$ defined by

$$
\beta(m) = \lim_{p \to p_c(m)+0} \frac{\log c(m, p, \infty)}{\log(p - p_c(m))},
$$

Within the mean-field approximation $\beta(\infty) = 1$.

The numerical results show that two regimes may be distinguished. In the small m regime—for $m \lesssim 10$—$p_c(m)$ and particularly $\beta(m)$ are close to their $m = 0$ values, whereas, in the large m regime—for $m \gtrsim 10$—they are close to their mean-field values.

The value found for $\beta(0)$ is the same as for directed percolation.[39] This result is in favor of a universal of critical behavior for one-dimensional cellular automata with one absorbing state.

In the case of long-range moves, the variations of $p_c(m)$ and β as foncions of m are strikingly different. While local and nonlocal site-exchange mixing processes lead to similar asymptotic behaviors for the stationary density $c(m, p, \infty)$ for small values of m when p is close to one, this is no more the case when p is close to p_c. When the stationary density is very small, the perturbation of the spatio-temporal pattern is much smaller for short-range moves than for long-range moves.

3   Automata network models

Automata networks are discrete dynamical systems, which may be defined as follows. Let $G = (V, E)$ be a graph, where V is a set of vertices and E a set of edges. Each edge joins two vertices not necessarily distinct. An automata network, defined on V, is a triple $(G, Q, \{f_i | i \in V\})$, where G is a graph on V, Q a finite set of states and $f_i : Q^{|U_i|} \to Q$ a mapping, called the local transition rule associated to vertex i. $U_i = \{j \in V | \{j, i\} \in E\}$ is the neighborhood of i, i.e., the set of vertices connected to i, and $|U_i|$ denotes the number of vertices belonging to U_i. The graph G is assumed to be locally finite, i.e., for all $i \in V$ $|U_i| < \infty$.

In all our models the set V is the two-dimensional torus Z_L^2, where Z_L is the set of integers modulo L.

3.1   Epidemics models

A vertex of Z_L^2 is either empty or occupied by either a *susceptible*, i.e., an individual who is not infected but who is capable of contracting the disease and become infective; or an *infective*, i e., an individual who is capable of transmitting the disease to susceptibles. The evolution results from the application of two subrules. The first subrule models infection, birth and death processes. It is a three-state cellular automaton rule applied *synchronously*. The second one specifies the motion of the individuals. It is applied *sequentially*. Both subrules are probabilistic and translation invariant, i.e., they do not depend upon the vertex i.

SIR MODEL

In an SIR model, based on disease status, the individuals are divided into three disjoint groups:

(S) the *susceptible* group, i.e. those individuals who are capable of contracting the disease and become infective;

(I) the *infective* group, i.e. those individuals who are capable of transmitting the disease to susceptibles; and

(R) the *removed* group, i.e. those individuals who have had the disease and are dead, or are isolated, or have recovered and are permanently immune.

The spread of the disease is governed by the following rules:

(i) Susceptibles become infective by contact, i.e. a susceptible may become infective with a probability p_i if, and only if, it is in the neighborhood of an infective. This hypothesis neglects incubation and latent periods. [n]

(ii) Infectives are removed with a probability p_r. This assumption states that removal is equally likely among infectives. In particular, it does not take into account the length of time the individual has been infective.

(iii) The time unit is the time step. During one time step, the two preceding rules are applied after the individuals have moved on the lattice.

(iv) Each individual performs an average of m tentative moves, either short- or long-range.

This model assumes that the population is closed. It ignores births, deaths by other causes, immigrations, or emigrations.

If we assume that the densities of the different groups of individuals are not space-dependent, the state of the system at time t is characterized by the respective densities $S(t)$, $I(t)$ and $R(t)$ of susceptible, infective, and removed individuals. $C = S(t) + I(t) + R(t)$ is the total density of individuals. If, moreover, we assume that correlations can be neglected, we obtain the mean-field equations:

$$S(t+1) = C - I(t) - R(t)$$

$$I(t+1) = I(t) + S(t)(1 - (1 - p_i I(t))^z) - p_r I(t)$$

$$R(t+1) = R(t) + p_r I(t)$$

z denotes the number of first neighbors. For Z_L^2, $z = 4$. From these equations, it follows that S is a positive nonincreasing function of t whereas R is positive nondecreasing. Therefore, the limits $S(\infty)$ and $R(\infty)$ exist. Since $I(t) = C - S(t) - R(t)$, it follows that $I(\infty)$ also exists and satisfies the relation

$$R(\infty) = R(\infty) + I(\infty),$$

which shows that $I(\infty) = 0$.

---

[n]During the latent period the individual who has been exposed to the disease is not yet infectious, whereas during the incubation period, the individual does not present symptoms but is infectious.

If the initial conditions are

$$R(0) = 0 \quad \text{and} \quad I(0) << S(0),$$

$I(1)$ is small, and we have

$$I(1) - I(0) = (zp_iS(0) - p_r)I(0) + O(I^2(0)).$$

Hence, according to the initial value of the density of susceptibles, we may distinguish two cases:

(i) If $S(0) < p_r/zp_i$ then $I(1) < I(0)$. Since S is a nonincreasing function of time, $I(t)$ goes monotonically to zero as t tends to ∞. That is, no epidemic occurs.

(ii) If $S(0) > p_r/zp_i$ then $I(1) > I(0)$. I(t) increases as long as the density of susceptibles $S(t)$ is greater than the threshold p_r/zp_i and then tends monotonically to zero.

This *threshold theorem* is the analogue of the Kermack-Mckendrick[18] theorem established using a model formulated in terms of differential equations.

The numerical results obtained for the automata network model,[40] which takes into account the local character of the interactions and the motion of the individuals, show that the mean-field results are qualitatively correct. They become exact in the limit $m \to \infty$. Note that to be able to observe cooperative phenomena, the toal density of individuals C has to be larger than the site percolation threshold, which is equal to 0.593.[41]

Sis model

In an Sis model, based on disease status, the individuals are divided into two disjoint groups:

(S) the *susceptible* group, i.e. those individuals who are capable of contracting the disease and become infective; and

(I) the *infective* group, i.e. those individuals who are capable of transmitting the disease to susceptibles.

The spread of the disease is governed by the following rules:

(i) Susceptibles become infective by contact, i.e. a susceptible may become infective witha probability p_i if, and only if, it is in the neighborhood o an infective. This hypothesis neglects incubation and latent periods.

(ii) Infectives recover with a probability p_r and become susceptible again. That is, at each time step, an infected individual either recover with a probability p_r or remains infective with a probability $1 - p_r$. The number of time steps T during which he remains infected is a geometric random variable, i.e. the probability $P(T = k)$ that T is equal to the positive integer k is equal to $p_r(1 - p - r)^{k-1}$, and we have

$$E(T) = 1/p_r \quad \text{and} \quad \text{Var}(T) = (1 - p_r)/p_r^2,$$

where $E(T)$ and $\text{Var}(T)$ denote respectively, the mean and the variance of T. This assumption states that the recovery is equally likely among infectives, it does take into account the length of time the individual has been infective.

(iii) The time unit is the time step. During one time step, the two preceding rules are applied after the individuals have moved on the lattice.

(iv) Each individual performs an average of m tentative moves, either short- or long-range.

For this model, if we assume that densities opf susceptibles and infectives do not depend on space, the state of the system at time t is characterized by the densities $S(t)$ and $I(t)$ of susceptibles and infectives, and the evolution equation of the density of infectives, within the framework of the mean-field approximation, is[42]

$$I(t + 1) = I(t) + S(t)\left(1 - \left(1 - p_i I(t)\right)^z\right) - p_r I(t),$$

where p_r denotes the probability per unit time for an infective to recover. Since the population is closed, the total density

$$C = S(t) + I(t)$$

is time-independent substituting $S(t) = C - I(t)$ in the expression of $I(t + 1)$ yields

$$I(t + 1) = I(t) + \left(C - I(t)\right)\left(1 - \left(1 - p_i I(t)\right)^z\right) - p_r I(t).$$

In the infinite-time limit, the stationary density of infectives $I(\infty)$ is such that

$$I(\infty) = (1 - p_r)I(\infty) + \left(C - I(\infty)\right)\left(1 - \left(1 - p_i I(\infty)\right)^z\right).$$

$I(\infty) = 0$ is always a solution of this equation. This value characterizes the disease-free state. It is a stable stationary state if, and only if, $zCp_i - p_r \leq 0$.

If $zCp_i - p_r > 0$, the stable stationary state is given by the unique positive solution of the previous equation. In this case, a nonzero fraction of the population is infected. The system is in the endemic state. For $zCp_i - p_r = 0$ the system, within the framework of the mean-field approximation, undergoes a transcritical bifurcation similar to a second order phase transition characterized by a nonnegative order parameter, whose role is played, in this model, by the stationary density of infected individuals $I(\infty)$. This threshold theorem is a well-known result for differential equation SIS models.[43]

It is easy to verify that, in the endemic state, when $zCp_i - p_r$ tends to zero from above, $I(\infty)$ goes continuously to zero as $zCp_i - p_r$. In the (p_i, p_r) parameter plane,

$$zCp_i - p_r = 0$$

is the equation of the second order phase transition line.

The numerical results obtained for the automata network model,[42] which takes into account the local character of the interactions and the motion of the individuals, show that the mean-field results are here again qualitatively correct. But, because it neglects correlations which play an essential role in the neighborhood of a second-order phase transition, the mean-field approximation cannot predict correctly the critical behavior of short-range interaction systems,[22] A detailed stiudy of the critical exponent β defined by

$$\beta = \lim_{p_i - p_i^c \to 0^+} \frac{\log I(m, \infty)}{\log(p_i - p_i^c)},$$

where $I(m, \infty)$ is the stationary value ($t = \infty$) of the density of infectives as a function of m, shows that for small m the system belong to the universality class of direct percolation [o] whereas for large m the critical behavior of the system becomes mean-field. As a function of m the system exhibits, therefore a crossover.[22]

GENERALIZED SIR MODEL

If, more generally, we assume that susceptibles and infectives may give birth to susceptibles at neighboring empty sites with respective probabilities b_s and b_i and that susceptibles may be removed with a probability d_s. The mean-field

---

[o]This result clearly shows how wrong is the assumption of homogeneous mixing.

volution equation read

$$S(t+1) = (1-d_s)S(t) + \left(1 - S(t) - I(t)\right) \times$$

$$f\left(b_s S(t) + b_i I(t)\right) - (1-d_s)S(t)f(p_i I(t)),$$

$$I(t+1) = (1-d_i)I(t) + (1-d_s)S(t)f(p_i I(t))$$

$$R(t+1) = R(t) + d_s S(t) + d_i I(t),$$

where $f(x) = 1 - (1-x)^z$.

In this model, the population is not closed. Due to birth processes, we obviously have

$$S(t+1) + I(t+1) + R(t+1) - S(t) - I(t)-$$

$$R(t) = (1 - S(t) - I(t))f(b_s S(t) + b_i I(t)).$$

This system exhibits 3 fixed points and a Hopf bifurcation. We will not enter in the details since the predator-prey model presented thereafter shows a similar behavior. The interested reader should refer to Boccara *et al.* (1994).[33]

4.2 A predator-prey model with pursuit and evasion

A vertex is either empty or occupied by either a *predator*, or a *prey*. In what follows, according to the process under consideration we will consider two different neighborhood. The *predation neighborhood* consists of the 4 nearest-neighbors of a given site, whereas the *pursuit and evasion neighborhood* consists of the 8 first and second nearest-neighbors.

The evolution of these two populations is governed by the following rules:

(i) A prey has a probability d_h to be captured and eaten by each predator in its predation neighborhood.

(ii) If there are no predators in its predation neighborhood, a prey has a probability b_h to give birth to a prey at an empty neighboring site.

(iii) After having eaten a prey, a predator has a probability b_p to give birth to a predator at the site previously occupied by the prey.

(iv) A predator has a probability d_p to die.

(v) Predators move to catch prey, and prey move to evade predators.

Predators (resp.prey) move in the direction of the highest (resp. lowest) loca prey density (resp. predator density) in their pursuit and evasion neighbor hood. If N is the total number of sites of Z_L^2, and c_p (resp. c_h) the predato (resp. prey) density, $m_p c_p N$ (resp. $m_h c_h N$) predators (resp. prey) are se quentially selected at random to perform a move. This sequential proces allows some individuals to move more than others. Since an individual ma only move to an empty site, the parameters m_p and m_h, which are positiv numbers, represent average numbers of *tentative* moves per individual durin a unit of time.

Rules i, ii, iii, and iv are applied *simultaneously*. Predation, birth an death processes are, therefore, modelled by a three-state two-dimensional ce lular automaton rule. Rule v is applied *sequentially*. At each time step, th evolution results from the application of the synchronous subrule followed b the sequential one.

The mean-field equations for this model are easily derived. Let $P(t)$ an $H(t)$ denote the densities at time t of, respectively, predators and prey. W have

$$P(t+1) = F_1(P(t), H(t))$$

$$= P(t) - d_p P(t) + b_p H(t) f(1, d_h P(t))$$

$$H(t+1) = F_2(P(t), H(t))$$

$$= H(t) - H(t) f(1, d_h P(t))$$

$$+ (1 - H(t) - P(t)) f(1 - P(t), b_h H(t)),$$

where the function f is defined by

$$f(p_1, p_2) = p_1^4 - (p_1 - p_2)^4.$$

The expression of $f(p_1, p_2)$ is derived as follows. If $p_2 = d_h P(t)$, then p_2 represents the probability that, at time t, a prey is eaten by a predator located at a specific site of the predation neighborhood. Then $1 - (1 - p_2)^4 = f(1, p_2)$ is the the probability that, at time t, a prey is eaten by a predator located at any of the 4 sites of the predation neighborhood. If $p_1 = 1 - P(t)$ and $p_2 = b_h H(t)$, then p_1 represents the probability that, at time t, a specifi site is not occupied by a predator and p_2 the probability that, at time t, a prey gives birth to a prey at specific site. Then $p_1^4 - (p_1 - p_2)^4 = f(p_1, p_2)$ i the probability that a prey gives birth to a prey at any of the 4 sites of the predation neighborhood is there are no predators in this neighborhood. Note that, within the framework of this approximation, the interaction terms are no ilinear as in most population dynamics models.[17,19,44] Nonbilinear interaction ave recently been shown to exhibit very different dynamic behaviors.[45]

The fixed points are the solutions of the equations

$$b_p H f(1, d_h P) - d_P = 0$$

$$(1 - H - P)f(1 - P, b_h H) - H f(1, d_h P) = 0.$$

These fixed points are stable if the absolute value of the eigenvalues $\lambda_1(P, H)$ and $\lambda_2(P, H)$ of the Jacobian matrix

$$J(P, H) = \begin{bmatrix} \dfrac{\partial F_1}{\partial P} & \dfrac{\partial F_1}{\partial H} \\[2mm] \dfrac{\partial F_2}{\partial P} & \dfrac{\partial F_2}{\partial H} \end{bmatrix}$$

are less than 1.

Studying the stability of these different fixed points, we find that:

$(0, 0)$ is always unstable if $b_h \neq 0$,

$(0, 1)$ is stable if $d_p - 4b_p d_h > 0$.

The expression of the Jacobian matrix $\beta J(P^*, H^*)$ corresponding to the non-trivial fixed point (P^*, H^*) is rather complicated and it is easier to study the stability of this fixed point numerically. An interesting feature of the model is that (P^*, H^*) may lose its stability, and a limit cycle becomes stable through a Hopf bifurcation.

We will not describe here the results of the numerical simulations. We shall only mention a rather interesting fact: for large m, the mean-field approximation still provides useful qualitative—although not exact—information on the general temporal behavior of such a system as a function of the different parameters. More precisely, for some parameter sets, the oscillatory behavior of the predator and prey populations, predicted by the mean-field approximation, is not observed for large lattices. However, we observed a *quasi-cyclic* behavior of the concentrations on a scale of the order of the mean displacements of the individuals.

Cyclic behaviors observed in population dynamics have received a variety of interpretations.[1] Our results suggest another possible explanation: approximate cyclic behaviors could result as a consequence of a *finite* habitat.

References

1. J.D. Murray, *Mathematical Biology*, (Springer-Verlag, Heidelberg 1989).

2. A.J. Lotka, *Elements of Physical Biology*, (Williams and Wilkins, Balti more 1925).

3. V. Volterra, *Variazioni e fluttuazioni del numero d'individui in speci animali conviventi*, R. Acc. dei Lincei vol. 6 **2**, 31–113 (1926).

4. V. Volterra *Elements of Mathematical Biology* (1956)

5. S.E. Kingsland, *Modeling Nature: Episodes in the History of Population Ecology* (University of Chicago Press, Chicago 1985).

6. G. R. Gavalas, *Nonlinear Differential Equations of Chemically Reacting Systems* (Springer-Verlag, New York 1968).

7. J.P. Finerty, *The Population Ecology of Cycles in Small Mammals* (Yale University Press, New Haven 1980).

8. E.T. Seton, *Life Histories of Northern Mammals: An Account of the Mammals of Manitoba* (Charles Scribner's Sons, New York 1909).

9. M. Begon, J.L. Harper and C.R. Townsend, *Ecology: Individuals, Populations and Communities* (Blackwell, Oxford 1986).

10. T.R. Malthus *An Essay on the Principle of Population, as it Affects the Future Improvement of Society, with Remarks on the Speculations of Mr Goodwin, M. Condorcet, and Other Writers* (Johnson, London 1798).

11. G.E. Hutchinson, *An Introduction to Population Ecology* (Yale University Press, New Haven 1978) pp. 11–18.

12. R.M. May, *Stability and Complexity in Model Ecosystems* (Princeton University Press, Princeton 1974) pp. 80–84.

13. E.C. Pielou, *Mathematical Ecology* (John Wiley, New York 1977) pp 91–95.

14. C.S. Holling, *The components of predation as revealed by a study of small-mammal predation of the european pine sawfly*, Can. Ent. **91**, 293–320 (1959).

15. C.S. Holling, *The functional response of predators to prey density and its role in mimicry and population*, Mem. Entomol. Soc. Can. **45**, 1–60 (1965).

16. J.T. Tanner *The stability and the intrinsic growth rates of prey and predator populations*, Ecology **56**, 855–867 (1975).

17. N. T. J. Bailey, *The Mathematical Theory of Infectious Diseases and its Applications*, (Charles Griffin, London 1975).

18. W.O. Kermack and A.G. McKendrick, *A contribution to the mathematical theory of epidemics*, Proc. Roy. Soc. **A 115**, 700–721 (1927).

19. P. Waltman, *Deterministic Threshold Models in the Theory of Epidemics*, (Springer-Verlag, Heidelberg 1974).

20. A. Källén, P. Arcuri and J.D. Murray, *A simple model for the spatial spread and control of rabies*, J. theor. Biol. **116**, 377–393 (1985).

21. J.D. Murray, E.A. Stanley and D.L. Brown, *On the spatial spread of rabies among foxes*, Proc. Roy. Soc. (London) **B 229**, 111–150 (1986).

22. N. Boccara, *Symétries Brisées*, (Hermann, Paris 1976).

23. P. Grassberger, *On the critical behavior of the general epidemic process and dynamical percolation*, Math. Biosci. **63**, 157–172 (1983).

24. J.L. Cardy and P. Grassberger, *Epidemic models and percolation*, J. Phys. A: Math. Gen. **18**, L267–L271 (1985).

25. G. McKay and N. Jan, *Forest fires as critical phenomena*, J. Phys. A: Math. Gen. **17**, L757–L760 (1984).

26. J.L. Cardy, *Field-theoretic formulation of an epidemic process with immunisation*, J. Phys. A: Math. Gen. **16**, L709–L712 (1983).

27. D. Mollison , *Spatial contact models for ecological and epidemic spread*, J. R. Statist. Soc. B **39**, 283–326 (1977).

28. E. Goles and S. Martínez, *Neural and Automata Networks*, (Kluwer, Dordrecht 1991).

29. K. Kuulasmaa, *The spatial general epidemic and locally dependent random graphs*, J. Appl. Prob. **19**, 745–758 (1982).

30. K. Kuulasmaa and S. Zachary, *On spatial general epidemic and bond percolation processes*, J. Appl. Prob. **21**, 911–914 (1984).

31. J.T. Cox and R. Durrett R., *Limit theorems for the spread of epidemics and forest fires*, Stoch. Proc. Appl. **30**, 171–191 (1988).

32. T.M. Liggett, *Interacting Particles Systems*, (Springer-Verlag, Heidelberg: 1985).

33. N. Boccara, K. Cheong and M. Oram , *A probabilistic automata network epidemic model with births and deaths exhibiting cyclic behaviour*, J. Phys. A: Math. Gen. **27**, 1585–1597 (1994).

34. S. Wolfram, *Statistical mechanics of cellular automata*, Rev. Mod. Phys. **55**, 601–644 (1983).

35. S. Wolfram, *Theory and Applications of Cellular Automata*, (World Scientific, Singapore 1986).

36. N. Boccara and M. Roger, *Some properties of local and nonlocal siteexchange deterministic cellular automata*, Intl. J. Mod. Phys. C **5**, 581–588 (1994).

37. N. Boccara, J. Nasser and M. Roger, *Particlelike structures and their interactions in spatio-temporal patterns generated by one-dimensional deterministic cellular-automaton rules*, Phys. Rev. **A 44**, 866–875 (1991).

38. N. Boccara and M. Roger, *Period-doubling route to chaos for a global variable of a probabilistic automata network*, J. Phys. A: Math. Gen. **25**, L1009–L1014 (1992).

39. J. Bease, *Series Expansions for the Directed-Bond Percolation Problem*,

J. Phys. C: Solid State Phys. **10**, 917–924 (1977).

40. N. Boccara and K. Cheong, *Automata network SIR models for the spread of infectious diseases in populations of moving individuals*, J. Phys. A Math. Gen. **25**, 2447–2461 (1992).

41. D. Stauffer, *Scaling theory of percolation clusters*, Physics Reports **54** 1–74 (1979).

42. N. Boccara and K. Cheong, *Critical behaviour of a probabilistic automata network SIS model for the spread of an infectious disease in a population of moving individuals*, J. Phys. A: Math. Gen. **26**, 3707–3717 (1993).

43. H.W. Hethcote, *Qualitative analyses of communicable disease models* Math. Biosci. **28**, 335-356 (1976).

44. R.M. Anderson and R.M. May, *Infectious Diseases of Humans, Dynamics and Control* (Oxford University Press, Oxford 1991).

45. H.W. Hethcote and P. van den Driessche, *Some epidemiological models with nonlinear incidence*, J. Math. Biol. **29**, 271–287 (1991).

46. N. Boccara, J. Nasser and M. Roger, *Annihilation of defects during the evolution of some one-dimensional class-3 deterministic cellular automata*, Europhys. Lett. **13**, 489–494 (1990).

47. N. Boccara, J. Nasser and M. Roger, *Critical behavior of a probabilistic local and nonlocal site-exchange cellular automaton*, Intl. J. Mod. Phys. C **5**, 537–545 (1994).

48. N. Boccara, E. Goles, S. Martínez and P. Picco (eds.), *Cellular Automata and Cooperative Phenomena. Proc. of a Workshop, Les Houches*, (Kluwer, Dordrecht 1993).

49. R. Bidaux, N. Boccara and H. Chaté, *Order of the Transition Versus Space Dimensionality in a Family of Cellular Automata*, Phys. Rev. A **39**, 3094–3105 (1989)

FORMAL NEURAL NETWORKS: AN INTRODUCTION TO SUPERVISED LEARNING AND A SELECTED BIBLIOGRAPHY

JEAN-PIERRE NADAL

Laboratoire de Physique Statistique de l'ENS[a]
Ecole Normale Supérieure,
24, rue Lhomond, F-75231 Paris Cedex 05, France.
nadal@lps.ens.fr
http://www.lps.ens.fr/˜nadal/

Abstract

This lecture is on the study of formal neural networks. The emphasis will be put on the supervised learning framework for feedforward architectures.

The brackets [B...] in the text refer to the sections of the Bibliography given at the end.

1 Introduction

Formal, or artificial, neural networks are processing units, simple models of the biological neurons. They are studied both in the context of neuroscience and neuropsychology, and in the one of computer science. There has been a large amount of work done in the 60's in order to understand the learning abilities of these formal neurons. New directions of research have emerge in the 80's, thanks to the efficacy of modern computers, allowing to performing heavy numerical simulations with complex neural architectures, and to new theoretical ideas and techniques, among which those derived from Statistical Physics.

In the study of formal neural networks one usually distinguishes two main types of learning paradigmes:

- Supervised learning: the desired output is given for a set of patterns. There are two sub-families:

  - learning by heart (that is realizing an associative memory)

  - learning a rule from example: the set of input-ouput pairs to be learned are a set of examples illustrating a rule. One expects the network to generalize, that is to give a correct output when a new (unlearned) pattern is presented.

- Unsupervised learning: no desired output is specified. The network "self organizes" as input patterns are sequentially presented.

Similarly, one may distinguish two main types of architectures:

- attractor neural networks (ANN), that is networks with a large amoun of feedback connections - possibly with every neuron receiving input: from every other neuron (Bibliography below, section [B.V]).

- feedforward networks made of layers, each layer receiving inputs from and only from the preceding layer. The simplest feedforward net is the perceptron (one input layer, one output layer, no "hidden" layer).

There are intermediate learning schemes, such as *reinforcement learning* and architectures, but it is convenient and useful to consider the above extreme cases. In this paper I will focus on the **supervised learning** framework for **feedforward networks** - for a more general introduction see e.g. the book by Hertz et al, 1991 (Bibliography, section [B.I.1]).

In section 2 I will review the main results on the performance of a *percep tron*, the simplest neural architecture. I will also show why this very simple architecture tells us something on the behavior of more complicated nets, such as multilayer networks. In section 3 I review the perceptron type algorithms that can be used either for an associative memory or for learning a rule from example. Eventually I indicate how more complicated architectures, in partic- ular multilayer networks, can be generated with the use of such algorithms.

2 Feedforward Neural Networks

2.1 The formal neuron and the perceptron

The simplest formal neuron is a binary element. In a network of formal neu- rons, neuron i ($i = 1, ..., N$) is in a state V_i taking the values 0 (neuron at rest) or 1 (neuron sending a spike). This neuron receives signals from other neurons. When a spike from neuron j arrives, it produces a change J_{ij} of the potential of the cell, where J_{ij} is the synaptic efficacy, or coupling, from neuron j to neuron i. At time t the local post synaptic potential is then

$$h_i = \sum_j J_{ij} V_j \tag{1}$$

where $V_j = 0, 1$ is the state of neuron j at that time t. The new state V_i is obtained from applying a threshold onto h:

$$V_i(t + 1) = \Theta(h_i(t) - \theta_i) \tag{2}$$

where $\Theta(h)$ is 1 for $h > 0$ and 0 otherwise.

In (almost) every neural network model a basic postulate is that learning is achieved through the modification of synaptic efficacies $\{J_{ij}\}$. This has been exploited during the 60's in the study of the simplest possible neural networks, in particular the *perceptron*. This network has an input layer directly connected to an output layer. The couplings between the two layers are adaptable elements (in the original design of the perceptron there is a preprocessing layer, but of fixed architecture and couplings: one can thus ignore it for all what follows). The simplest perceptron has only one output, a binary unit, V, with a (possibly) large number N of inputs $\{\xi_j, j = 1, ..., N\}$ which may have continuous or discrete states:

$$V = \Theta(\sum_{j=1}^{N} J_j \xi_j - \theta) \tag{3}$$

In a supervised learning task, one is given a set X of p input patterns,

$$X = \{\vec{\xi}^{\mu}, \mu = 1, ..., p\} \tag{4}$$

and the set of the corresponding desired outputs,

$$\vec{\tau} = (\tau^{\mu} = 0, 1, \ \mu = 1, ..., p)$$

which have to be learned by the perceptron. For a given choice of the couplings, the output V^{μ} when the μth pattern is presented is given by:

$$V^{\mu} = \Theta(\sum_{j=1}^{N} J_j \xi_j^{\mu} - \theta) \tag{5}$$

Learning means choosing (computing. adapting) the couplings and the threshold in such a way that the desired output configuration $\vec{\tau}$ is equal - or as close as possible - to the actual output $\vec{V} = (V^{\mu} = 0, 1, \ \mu = 1, ..., p)$.

Simple variant of the formal neuron are also commonly studied. In particular neurons with a continuous state obtained by applying a smooth *transfert function f* instead of the hard thresholding:

$$V = f(h - \theta) \tag{6}$$

Such model neurons are used in many applications of artificial neural networks (e.g. for function approximation tasks, see the Bibliography, [B.VII.3]), with the popular *backprop* algorithm (see [B.III]). More complicated models are also considered in neuroscience (see e.g. Arbib 1995, [B.I.2], and [B.V.3]). In this paper I will however restrict to the case of the binary formal neuron (2,3). In the next section I consider the ability of the perceptron (3) to learn.

2.2   The geometrical approach

The storage capacity of the percetron has been obtained from geometrical arguments (Cover 1965). One considers the space of couplings: $\vec{J} = \{J_j, j = 1, ..., N\}$ is a point in an N dimensional space. Then each pattern μ defines an hyperplane in this space, and the output V^μ is 1 or 0 depending on which side of the hyperplane lies the point $\vec{J}$. Hence the p hyperplanes devide the space of couplings in domains (figure [1]), each domain being associated to one specific set $\vec{V} = \{V^1, ..., V^p\}$ of outputs. Let us call $\Delta(X)$ the number of domains. Since each V^μ is either 0 or 1, there is at most 2^p different output configurations $\vec{V}$, that is

$$\Delta(X) \leq 2^p. \tag{7}$$

If the patterns are "in general position" (that is every subset of at most N patterns are linearly independant), then a remarkable result is that $\Delta(X)$ is in fact independant of X and a function only of p and N (Cover 1965):

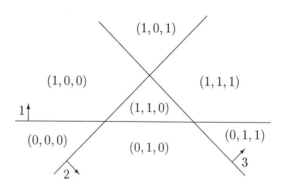

Figure 1: Partition of $\vec{J}$ space in domains. Here $p = 3$ patterns in $N = 2$ dimensions define 7 domains. For each pattern the arrow points towards the half space of the J's producing an output 1 for this pattern. The resulting output configuration $\vec{V} = (V^\mu, \mu = 1, 2, 3)$, is given inside each domain. The output configuration $\vec{V} = (0, 0, 1)$ is not realized.

$$\Delta = \sum_{k=0}^{\min p, N} C_p^k \tag{8}$$

where $C_p^k = \frac{p!}{k!\,(p-k)!}$. In particular

$$\Delta = \begin{cases} 2^p & \text{if } p \leq N \\ < 2^p & \text{if } p > N \end{cases} \tag{9}$$

This means that $N+1$ is the "Vapnik-Chervonenkis dimension" (Vapnik 1995, [B.VII.2]) of the perceptron ($N+1$ is the first value of p for which Δ is smaller than 2^p):

$$d_{VC} = N + 1 \tag{10}$$

If the task is to learn a rule from example, the VC dimension plays a crucial role: generalization will occur if the number of examples p is large compared to d_{VC} (Vapnik 1995). Another important parameter is the asymptotic capacity α_c. In the large N limit, for a fixed ratio

$$\alpha \equiv \frac{p}{N} \tag{11}$$

the asymptotic behaviour of $\Delta/2^p$ shows that the fraction of output configurations which are not realized remains vanishingly small for α not greater than 2, and becomes 1 above (Cover 1965, Gardner 1987, 1988, [B.VI]), and this defines a critical capacity,

$$\alpha_c = 2. \tag{12}$$

In fact $C \equiv \ln \Delta$ has an important meaning: it has been shown by G. Toulouse (see Brunel et al 1992, [B.VI.2]) that, for any α, C is the maximal amount of (Shannon) information that can be stored in the synapses. Above α_c, one can relate the minimal number of errors that can be achieved to this information capacity. This result can be seen (Nadal 1994, [B.I.1]) as an application of a fundamental theorem in information theory (see e.g. Blahut 1988 [B.I.2]) giving the smallest possible error rate that can be achieved by a communication channel of (Shannon) capacity C.

2.3 Multilayer perceptrons: an optimal upper bound

The preceding results for the perceptron appear to be also useful when considering more complex architectures - in fact any learning machine with a binary output. For a general learning machine, the VC dimension and the number of domains are defined as above : $\Delta(X)$ is the number of different possible output configurations $\vec{V}$. In general it will depend on the choice of X (and not only on p and N as for the perceptron). However one can consider its maximal value over all possible choices of X:

$$\Delta_m \equiv \max_X \Delta(X) \tag{13}$$

This maximal value Δ_m is equal to 2^p for p up to some number called the VC (Vapnik-Chervonenkis) dimension, d_{VC} (possibly infinite), and is strictly smaller above. As mentioned above for the perceptron, generalization is guaranteed for p much larger than the VC dimension. It has been shown the remarkable result that Δ_m is bounded above by $\sum_{k=0}^{\min p, d_{VC}} C_p^k$ (Vapnik 1995). Among all the learning machines with a given value d of the VC dimension, the perceptron with $N = d - 1$ inputs is a particular one for which this bound is an equality. This implies that we can consider the learning performance of the perceptron with $N = d - 1$ inputs as an approximation for the learning performance of any other learning machine for which d is the VC dimension.

2.4 Statistical physics approach to learning

In 1987 Elizabeth Gardner introduced a statistical physics approach to the study of learning ([B.VI.2]). She considered a *measure* in the space of couplings, so that it is possible to ask for the number (or the fraction) of couplings that effectively learn a set of pattern - hence one computes the typical *volume* of the domains considered above.

From that approach, using the techniques developed for the study of spin-glass models (see Mézard et al 1987, [B.I.3]), one gets the storage capacity of the perceptron under various conditions (unbiased or biased patterns, continuous or discrete couplings,...; the critical capacity $\alpha_c = 2$ corresponding to the particular case of continuous couplings and unbiased patterns). One gets also the typical behavior of a network taken at random among all the networks which have learned the same set of patterns. Moreover this approach has been adapted to the study of generalization, that is to the learning of a rule from example (Tishby et al 1989, [B.VII.1]).

3 Algorithms: the perceptron and beyond

3.1 Learning algorithms for the perceptron

We know that a perceptron can learn at most $2N$ associations, but is it possible to find one set of couplings that realize this learning? The *perceptron algorithm* proposed by Rosenblatt (see Minsky and Papert 1988) allows precisely to find a solution. A remarkable fact is that it is possible to proove that the algorithm will converge in a finite amount of time (whenever a solution does exists). This algorithm is very simple: it consists in taking a pattern at random, checking wether the current couplings give a correct output; if not, one performs a learning step, with a Hebbian rule (if pattern μ is being tested, each coupling J_j is increased if the input ξ_j^μ and the desired output τ^μ have the same sign, and decreased otherwise). This procedure is repeated until convergence. In practice one has to let run the algorithm a given, arbitrarily chosen, amount of time, since one does not know in advance wether at least one solution exists.

Many variant of the basic algorithm have been proposed (see [B.II]) in order to find couplings satisfying some specific properties. Some of them have been developped in the context of the modeling of associative memories, see [B.V], in which perceptron type algorithms can also be used, eventhough the architecture of the network is very different.

In particular several algorithms - the "minover" (Krauth and Mézard 1987), the "adatron" (Anlauf and Biehl 1989) and the "optimal margin classifier" (Boser et al 1992) - allow to find a specific solution which is known to produce very good - although not necessarily optimal - performance in both associative properties (Krauth and Mézard 1987) and generalization (Vapnik 1995). This particular solution has been generalized to more complex feedforward architectures, leading to the concept of *Support Vector Machines* (Vapnik 1995).

Other related algorithms have been designed for the case where the desired associations are not learnable exactly. There are various algorithms which tend to find couplings such that the number of errors will be as small as possible. In particular, the "pocket" algorithm (Gallant 1986, [B.III]) is a variant of the perceptron algorithm which guarantees to find a solution with the smallest possible number of errors - provided one lets run the algorithm long enough...

3.2   Constructive algorithms

In most practical applications, where one wants to find a rule hidden behind a set of examples, an architecture more complicated thant the one of the perceptron is required. The most standard approach is to choose a layered network with an a priori chosen number of hidden layers, and to let run the *backpropagation* algorithm (see e.g. Hertz et al [B.I.1], and [B.III]). There exists however alternatives to this method: one can also "learn" the architecture. Since 1986 there exists a family of constructive algorithms, which adds units until the desired result is obtained (see [B.IV]). Most of these algorithms are based on perceptron learning. I give here one example, the "Neural Tree" algorithm (Sirat and Nadal 1990, Golea and Marchand 1990, [B.IV.2]) (also called the "upstart" algorithm in the slightly different version of M. Frean).

Given the "training set", a set of p input patterns with their class 0 or 1 (their desired output $\tau = 0$ or 1), one starts by trying a perceptron algorithm in order to learn the p associations (pattern-class): in case theses associations were learnable by a perceptron, the algorithm will give one solution, and the

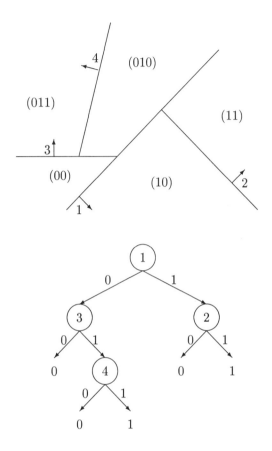

Figure 2: A Neural Tree. Above: partition of the input space by a Neural Tree. Below: The functional tree architecture. (*Adapted from Nadal 1994, [B.I.1]*)

problem is solved. If not (in practice if no solution has been found after some given amount of time), then one keeps the couplings given by the algorithm (or the *pocket* solution (Gallant 1986), that is the set of couplings with the least number of errors). These couplings define our first neuron. They define an hyperplane which cuts the input space into two domains (figure 2): input patterns on one side have a $V_1 = 1$ output, patterns on the other side have a $V_1 = 0$ output. At least one of these domains contains a mixture of patterns of the two classes. We will say that such a domain is *unpure*, a *pure* domain being one which contains patterns of a same class. The goal of the algorithm is to end up with a partition of the input space in pure domains. One considers each unpure domain separately. For a given (unpure) domain, one lets run a perceptron algorithm trying to separate the patterns according to the class they belong to. This leads to a new unit, defining an hyperplane which cuts the domain into two new domains. This procedure is repeated until every domain is pure. On figure 2 five domains have been generated.

One should note that every neuron that has been built is receiving connections from (and only from) the input units. The tree is functional: consider for example the neural tree of figure 2; to read the class of a new pattern, one looks at the output of the first neuron. Depending on its value, 1 or 0, one reads the output of neuron 2 or 3. In the first case, the output of neuron 2 gives the class. In the second case, one reads the output of neuron 4 which gives the class.

One should note also that the perceptron algorithm can be replaced by any learning algorithm (for the perceptron architecture) that one finds convenient. Most importantly, this algorithm can be easily adapted to multiclass problems (Sirat and Nadal 1990), that is when the desired output can take more than two values: in the final Neural Tree, each domain will contain patterns of a same class.

The best performances on generalization may not be obtained when every example of the training set is correctly learned: this may be due to noisy data, or to the rule itself which might be stochastic. But with a constructive algorithm one can always add units until every output is equal to the desired output. Hence it is likely that the net will in fact "learn by heart" every example and will not generalize very well. Indeed, one has to stop the growth of the tree when generalization, as measured by the number of correct answers on a test set, starts to decrease (d'Alché and Nadal 1995). Such strategy can be applied locally, that is at each leave of the current tree. This is an advantage of this algorithm: the input space is partitioned in a way that reflects the local density of data, so that one has a good control on the quality of generalization.

4 Conclusion

I have presented some results on supervised learning in feedforward networks made of formal neurons: learning ability of the perceptron, learning algorithms for the perceptron and more complex feedforward architectures, with emphasis on the constructive approach. The interested reader will find below a selection of general and specific references.

Bibliography

Alphabetical order within each subject.

B.I. General references

B.I.1. Books and review papers on formal neural networks

- Amit D. J., *Modeling Brain Function*. Cambridge University Press, 1989.

- Guyon I., Neural networks and applications tutorial. In K. M. Decker, editor, *Parallel Architectures ans Applications*. Computer Physics Reports, vol. 207, num. 3-5, North-Holland, 1991.

- Hertz J., Krogh A., and Palmer R. G., *Introduction to the Theory of Neural Computation*. Addison-Wesley, Cambridge MA, 1991.

- Kohonen T.O., *Self-organization and associative memory*. Springer, Berlin, 1984.

- Minsky M.L. and Papert S.A., *Perceptrons*. M.I.T. Press, Cambridge MA, 1988.

- Nadal J.-P., Formal neural networks: from supervised to unsupervised learning, in *Cellular Automata, dynamical systems and neural networks* E. Goles & Martinez editors (Kluwer 1994) pp.147-166.

- Peretto P., *An introduction to the modeling of neural networks*. Cambridge University Press, 1992.

B.I.2. Beyond, and behind, formal neural networks

- Arbib M. A. Editor, *The Handbook of Brain Theory and Neural Networks*. Bradford Books/The MIT Press, 1995.

- Blahut, R. E., *Principles and Practice of Information Theory*, Addison Wesley, 1988.

- Grassberger P. and Nadal J.-P. Editors, *From Statistical Physics To Statistical Inference and Back*, NATO ASI Series C, Vol. 428. Kluwer Acad. Pub., Dordrecht, 1994.

- Gutfreund H. and Toulouse G., *Biology and Computation: a physicist's choice*. World Scientific Pub., Singapore, 1994.

- Widrow B. and Stearns S. D., *Adaptive Signal Processing*. Prentice-Hall, 1985.

B.I.3. More specifically for physicists

- Geszti T., *Physical Models of Neural Networks*. World Scientific Pub., Singapore, 1989.

- Mézard M., Parisi G., and Virasoro M., *Spin Glass Theory and Beyond*. World Scientific Pub., Singapore, 1987.

- Sompolinsky H., Statistical mechanics of neural networks. *Physics Today*, 40:70–80, December 1988.

B.II. Perceptron type algorithms

- Abbott L. F., Learning in neural network memories. *NETWORK*, 1:105–22, 1990.

- Anlauf J. K. and Biehl M., The adatron: an adaptative perceptron algorithm. *Europhys. Lett.*, 10:687, 1989.

- Boser B., Guyon I., and Vapnik V., An algorithm for optimal margin classifiers. In *Proceedings of the ACM workshop on Computational Learning Theory*, Pittsburgh, July 1992, 1992.

- Gallant S.I., Optimal linear discriminants. In *Proceedings of the 8 th . Int. Conf. on Pattern Recognition*, pages 849–52, Paris 27-31 October 1986, 1987. IEEE Computer Soc. Press, Washington D.C.

- Krauth W. and Mézard M., Learning algorithms with optimal stability in neural networks. *J. Phys. A: Math. and Gen.*, 20:L745, 1987.

- Minsky M.L. and Papert S.A., *Perceptrons*. M.I.T. Press, Cambridge MA, 1988.

B.III. Backpropagation

- Le Cun Y., A learning scheme for asymmetric threshold networks. In *Proceedings of Cognitiva 85*, pages 599–604, Paris, France, 1985. CESTA-AFCET.

- Rumelhart D. E., Hinton G. E., and Williams R. J., Learning internal representations by error propagation. In McClelland J. L. Rumelhart D. E. and the PDP research group, editors, *Parallel distributed processing: Explorations in the microstructure of cognition, volume I*, pages 318–362. Bradford Books, Cambridge MA, 1986.

B.IV. Constructive Algorithms

B.IV.1. Building multilayer perceptrons

- Bichsel M. and Seitz P., Minimum class entropy: a maximum information approach to layered networks. *Neural Network*, 2:133–41, 1989.

- Gallant S.I., Three constructive algorithms for network learning. In *Proc. 8th Ann Conf of Cognitive Science Soc*, pages 652–60, Amherst, MA 15-17 august 1986, 1986.

- Grossman T., Meir R., and Domany E., Learning by choice of internal representations. *Complex Systems*, 2:555, 1988.

- Mézard M. and Nadal J.-P., Learning in feedforward layered networks: the tiling algorithm. *J. Phys. A: Math. and Gen.*, 22:2191–203, 1989.

- Rujan P. and Marchand M., Learning by minimizing resources in neural networks. *Complex Systems*, 3:229–42, 1989.

- Torres Moreno J. M. and Gordon M., Efficient adaptive learning for classification tasks with binary units Submitted to *Neural Computation*, preprint 1997

B.IV.2. Hierarchical architectures

- d'Alché Buc F., Zwierski D., and Nadal J.-P., Trio learning: a new strategy for building hybrid trees. *to appear in Int. Journ. of Neur. Syst.*, 1994.

- d'Alché Buc F. and Nadal J.-P., Asymptotic performance of a constructive algorithm. *Neural Processing Letters*, 1:1–4, 1995.

- Frean M., The upstart algorithm: a method for constructing and training feedforward neural networks. *Neural Comp.*, 2:198–209, 1990.

- Golea M. and Marchand M., A growth algorithm for neural network decision trees. *Europhys. Lett.*, 12:105–10, 1990.

- Jordan M. I. and Jacobs R. A., Hierarchical mixtures of experts and the em algorithm. *Neural Comp.*, 6:181–214, 1994.

- Knerr S., Personnaz L., and Dreyfus G., Single layer learning revisited: a stepwise procedure for building and training a neural network. In Fogelman F. and Hérault J., editors, *Proc. NATO workshop Les Arcs 1989.* Springer, 1990.

- Sirat J. A. and Nadal J.-P., Neural trees: a new tool for classification. *NETWORK*, 1:423–38, 1990.

B.V. Associative Memory

B.V.1 A precursor: D. O. Hebb

- Hebb D. O., *The Organization of Behavior: a neurophysiological study.* John Wiley, New-York, 1949.

B.V.2 Hopfield type models

- Amit D. J., Gutfreund H., and Sompolimsky H., Storing an infinite number of patterns in a spin-glass model of neural networks. *Phys. Rev. Lett.*, 55:1530–1533, 1985.

- Hopfield J.J., Neural networks as physical systems with emergent computational abilities. *Proc. Natl. Acad. Sci. USA*, 79:2554–58, 1982.

- Meunier C. and Nadal J.-P., Sparsely coded neural networks. In Arbib M., editor, *The Handbook of Brain Theory and Neural Networks.* Bradford Books/The MIT Press, 1995, pp. 899-901.

B.V.3 More realistic networks

- Amit D. J., The hebbian paradigm reintegrated: local reverberations as internal representations. *BBS*, 1994
http://www.cogsci.soton.ac.uk/bbs/Archive/bbs.amit.html.

- Amit D. J. and Brunel N., Model of global spontaneous activity and local structured delay activity during delay periods in the cerebral cortex. *Cerebral Cortex*, 7:237–252, 1997.

B.VI. Storage capacity

B.VI.1. Geometrical approach

- Cover T. M., Geometrical and statistical properties of systems of linear inequalities with applications in pattern recognition, *IEEE Trans. Electron. Comput.*, 14:326, 1965.

B.VI.2. Statistical Mechanics approach

- Brunel N., Nadal J.-P. and Toulouse G., Information Capacity of a Perceptron, *J. Phys. A* 25:5017-5037, 1992.

- Gardner E., Maximum storage capacity in neural networks. *J. Phys. I (France)*, 48:741–755, 1987.

- Gardner E., The space of interactions in neural networks models. *J. Phys. A: Math. and Gen.*, 21:257, 1988.

- Krauth W. and Mézard M., Storage capacity of memory networks with binary couplings. *J. Phys. I (France)*, 50:3057, 1989.

- Reimann P. and Van den Broeck C., *Phys. Rev. E* 53:3989, 1996.

B.VII. Learning from examples

B.VII.1. Statistical Mechanics approach

- Levin E., Tishby N., and Solla S. A. A statistical approach to learning and generalization in layered networks. In *1989 workshop on computational learning theory, COLT'89*, 1990.

- Tishby N., Levin E., and Solla S. Consistent inference of probabilities in layered networks: predictions and generalization. In *Proceedings of the IJCNN International Joint Conference on Neural Networks (Washington D.C., 1989)*, volume 2, pages 403–409. IEEE, 1989.

- T. Watkin, A. Rau, and M. Biehl. The statistical mechanics of learning a rule. *Review of Modern Physics*, 65:499–556, 1993.

B.VII.2. Structural Risk Minimization (Vapnik)

- Vapnik V. *Estimation of Dependences Based on Empirical Data*. Springer Series in Statistics. Springer, New-York, 1982.

- Vapnik V. *The Nature of Statistical Learning Theory*. Springer Series in Statistics. Springer, New-York, 1995.

B.VII.3. Function Approximation

- Barron A. R. Universal approximation bounds for superpositions of a sigmoidal function. *IEEE Trans. I. T.*, 39, 1993.

- Cybenko G. Approximations by superpositions of a sigmoidal function. *Math. Contr. Signals Syst.*, 2:303–314, 1989.

- Hornik K., Stinchcombe M., and White H. Multilayer feedforward networks are universal approximators. *Neural Networks*, 2:359–366, 1991.

- Weigend A. S. and Gershenfeld N. A., Time series prediction, *Addison-Wesley*, 1994

DYNAMICAL SYSTEMS, QUALITATIVE THEORY, AND SIMULATION

ARKADY PIKOVSKY

Department of Physics
Potsdam University, Potsdam, Germany

1 What is dynamical approach?

The Webster defines dynamics as the pattern of history of growth, change, and development in any field.

When one speaks about "dynamical systems", or "nonlinear dynamics", or "dynamical properties", it means that the point of major interest is the development in time. Different objects can be considered – wheather, stock markets, positions of stars, number of cancer cells – and to all of them the dynamical approach can be applied.

In a more narrow sense, one tries to describe the development in time scientifically, i.e. to derive it from some laws of motion, which govern the evolution. In many cases one knows these laws (like in Newtonian mechanics), in other cases the laws can be guessed (like in cell growth), but in many situations the time evolution is subject to uncontrollable external influences (like stock market). At this point it is useful to distinguish deterministic systems - governed by some laws (although, may be, not known to an observer); and stochastic systems, where random forces act. We consider only the former case, when one has deterministic dynamical process.

Dealing with a dynamical process, one can state different problems. Considering wheather variations, the main problem is forecasting: using the laws (which are in reality extremely complex equations) one tries to calculate the development of the system in time to find future wheather, and in fact it is possible to do this for a couple of days. It is clear, that for such calculations one has not only to know the laws, but also the present state of the system. In many cases, however, a particular evoilution history is not of large interest, but some general questions on the time development are important: e.g. will the Moon (or, more practical, a sattelyte) fall on the Earth; or will the number of cancer cells grow or decrease, or may be oscillate. Exactly to questions of this kind the qualiative theory of dynamical systems is devoted. There are two major classes of dynamical systems: with continuous and discrete time. Continuous-time systems are represented by ordinary differential equations. In general, there is no hope hope to solve a system of nonlinear differential equations analytically, and the qualitative theory only predicts what type of the dynamics can be observed for different values of parameters and initial conditions. The same can be said about discrete dynamical systems – maps.

The lectures are organized as follows: we start from the simplest (low-dimensional) dynamical systems, and then proceed to higher dimensions and more complex regimes. We also briefly discuss influence of noise, dynamics of distributed systems, and numerical methods.

2 One-dimensional dynamics

The simplest dynamical system is given by a one-dimensional differential equation

$$\frac{dx}{dt} = f(x) \tag{1}$$

The analysis of this system is straightforward: one should draw the function $f(x)$. The zeros of this function are fixed points of the dynamical system: if $f(x^*) = 0$, then $x = x^*$ is a constant in time solution of Eq. (1). The second step is finding the stability of the fixed points. Roughly speaking, a fixed point is stable, if any tarjectory starting from a small vicinity of this point tends to this point. Otherwise a fixed point is unstable. Stability condition for Eq. (1) can be derived just from the geometrical representation of the dynamics. Note that x grows if f is positive, and x decreases if f is negative, so one can put arrows on the line x. Thus, fixed points where f changes from positive to negative values, i.e. $f' < 0$, are stable, and those where f changes from negative to positive value, i.e. $f' > 0$, are unstable. Let us assume that the trajectories of the dynamical system (1) do not escape to infinity, i.e. $f > 0$ at $x \to -\infty$ and $f < 0$ at $x \to \infty$. Then, there exists at least one stable fixed point, and starting from almost any initial condition, the trajectory ends in one of the stable fixed points. The dynamics can be divided into two stages:

1. transients: motion from the initial point to the stable fixed point;

2. steady state: the trajectory stays at the stable fixed point.

This division is a general property of nonlinear dissipative dynamical systems: after initial transients something stationary appears and remains forever. The object in the phase space, corresponding to the stationary state, is called *attractor*. For the one-dimensional dynamical system the only possible attractor is a stable fixed point. Note, that a system can have more than one stable fixed point, and, more general, more than one attractor. Each attractor has its basin of attraction – a neughbourhood, starting from which trajectories are approaching this attractor. In the nonlinear dynamics one is usually interested

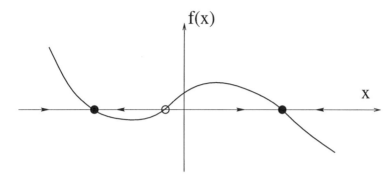

Figure 1: A sketch of one-dimensional dynamics. Filled and empty sircles mark stable and unstable fixed points.

in the behavior at large times, therefore one needs to study attractors and their basins.

Consider an example of one-dimensional dynamical system:

$$\frac{du}{dt} = au - u^3 + b \tag{2}$$

Here two parameters a, b govern the system, depending on their values different dynamics is observed:

- In region I only one attractor (a stable fixed point) exists, whose basin of attraction is the whole phase space $-\infty < u < \infty$.

- In region II there are 3 fixed points: 2 stable $u_{1,2}^s$ and 1 unstable u^u. The basin of the attractor u_1^s is a halfplane $-\infty < u < u^u$, and the basin of attraction of u_1^s is a halfplane $u^u < u < \infty$. This situation is called *bistability*, a more general case is called *multistability*.

How the transition from region I to region II occurs? If we draw the function $ax - x^3 + b$ on the both sides of the transition line, we can see that as the parameters moove from the region II towards the border, a stable fixed point comes closer to the unstable one, and at the border both collide and disappear. In general, such qualitative changes of the dynamical objects (fixed points, attractors) are called *bifurcations*. The transition we described is a particular bifurcation when a stable and an unstable fixed points collide and disappear, it is called saddle-node bifurcation. Correspondingly, the border lines between qualitatively different regimes on the plane (or, more generally, space) of parameters are called *bifurcation lines*. Essentially, investigation of a

116

Figure 2: Dynamics of the equation (2): the bifurcation diagram, the saddle-node bifurcation, and the hysteresis.

dynamical model usually means finding attractors, their basins, and drawing a bifurcation diagram on the parameter plane. After doing this one can predict qualitative behavior of the system: what happens with trajectories for given parameter values, and what happens when the parameters are changed. For example, from the bifurcation diagram for the equation (2) one can describe what happens if the parameter b is changed: as a bifurcation line is crossed, a jump occurs from one attractor to another, and these jumps happen at different places depending on whether b is increasing or decreasing; this phenomenon is called *hysteresis*.

3 Two-dimensional dynamics

The dynamics of two-dimensional systems

$$\frac{d^2x}{dt^2} = f(x,y), \qquad \frac{d^2y}{dt^2} = g(x,y) \qquad (3)$$

includes, of course, all the possibilities of one-dimensional dynamics, but additionally some new features. The most important is a possibility to have a new type of attractor: a closed curve on the phase plane (x,y), it is called *limit cycle*. Like a fixed point, a limit cycle can be stable or unstable. A stable limit cycle describes periodic self-sustained oscillations in the dynamical system.

Example 1 (A mathematical model of the spread of gonorrhea[6]) The assumptions for the construction of the model are as follows:

1. An individual becomes infective immediately after contracting gonorrhea.

2. The promiscuous population can be split into two groups, susceptibles and infectives.

3. Male and female infectives are cured at rates α_1 and α_2, respectively, proportional to their total number. Usually $\alpha_1 > \alpha_2$.

4. New infectives are added to the male (female) populations at rates β_1 (β_2) proportional to the total number of male (female) susceptibles and female (male) infectives.

5. The total number of promiscuous males and females remain at constant levels c_1 and c_2, respectively.

Denoting the number of infective males and females by m and f, we get the following system:

$$\frac{dm}{dt} = -\alpha_1 m + \beta_1 (c_1 - m)f, \qquad \frac{df}{dt} = -\alpha_2 f + \beta_2 (c_2 - f)m \qquad (4)$$

Let us find fixed points of the system:

$$m_0 = f_0 = 0, \qquad m_1 = \frac{\beta_1 \beta_2 c_1 c_2 - \alpha_1 \alpha_2}{\beta_2 \beta_1 c_2 + \alpha_1 \beta_2}, \qquad f_1 = \frac{\beta_1 \beta_2 c_1 c_2 - \alpha_1 \alpha_2}{\beta_2 \beta_1 c_1 + \alpha_2 \beta_1}$$

The bifurcation point is $\beta_1 \beta_2 c_1 c_2 - \alpha_1 \alpha_2 = 0$, here the stable fixed point m_0, f_0 becomes unstable, and the stable fixed point m_1, f_1 appears. The bifurcation condition can be rewritten as

$$\left(\frac{\beta_1 c_1}{\alpha_2} \right) \left(\frac{\beta_2 c_2}{\alpha_1} \right) = 1.$$

The ratio $\beta_{1,2} c_{1,2} / \alpha_{2,1}$ can be interpreted as the number of males (females) that a female (male) infective contacts during her (his) infectious period, if every male (female) is susceptible. In 1973 in Rhode Island, the numbers named were 1.15 and 0.98 (the real numbers are higher). Thus, gonorrhea ultimately approaches a nonzero steady state m_1, f_1.

The appearance of a limit cycle usually happens via the Hopf bifurcation. Befor the bifurcation a stable fixed point exists, with a certain basin. The transients are not monotonous, but have a form of damped oscillations. As the bifurcation point is approached, the damping rate decreases and the transient time increases. Exactly at the transition point the transient time tends to infinity. After the bifurcation point a small limit cycle appears, with the amplitude growing roughly as square root of the distance to the bifurcation point.

Example 2 (A preadtor-prey system.) Let x and y represent the number of preys and predators, respectively. We assume that the birth rate in the prey is c_1, and the death is caused by the overpopulation of the preys (the term $-c_2 x^2$) and by the predators (the term proportional to c_3). The birth rate of predators is c_5, and their death rate is inverse proportional to the population of prey. The equations are

$$\frac{dx}{dt} = c_1 x - c_2 x^2 - \frac{c_3 xy}{x + c_4}, \qquad \frac{dy}{dt} = yc_5 \left(1 - \frac{c_6 y}{x} \right).$$

Fixing $c_1 = c_2 = c_6 = 1$, $c_3 = 1.5$, $c_4 = 0.2$ and changing c_5, one observes a bifurcation from a stable fixed point to a limit cycle (Fig. 3).

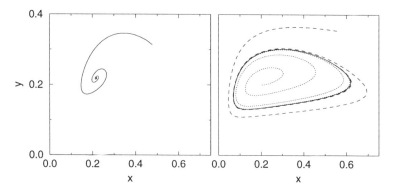

Figure 3: The dynamics in the predator-prey system for $c_5 = 0.4$ (stable fixed point) and $c_5 = 0.1$ (limit cycle)

In general, it is rather hard, if not impossible, to investigate a dynamical system analytically. Some methods exist, which are applicable if the system has special properties. For example, near the Hopf bifurcation point a method of averaging can be used, based on the smallness of the solution. I describe below another method, which is applicable to so-called relaxational systems. An example is the Hodgkin-Huxley model of the nerve excitation. This model has four variables, and is rather tedious. Its simplification has been suggested by FitzHugh and Nagumo.

Example 3 (The FitzHugh-Nagumo model.)

$$\varepsilon \frac{dx}{dt} = -x^3 + x - y = P(x, y) \tag{5}$$

$$\frac{dy}{dt} = x - by + a = Q(x, y) \tag{6}$$

Here x describes a membrane potential, and y the ion potential. The parameter ε is small. To investigate the system, let us draw the lines $P(x, y) = 0$ and $Q(x, y) = 0$ on the phase plane (x, y) (Fig. 4). The time evolution proceeds in two steps

1. Fast motion: during time $O(\varepsilon)$ the variable x goes to a line $P = 0$, while y remains a constant.

2. Slow motion: the variable y slowly evolves, while $P(x, y) \approx 0$.

Note that the fast motion is described by the first-order equation (5), which looks like (2). This equation can have, depending on y, one or three fixed points, and a stable fixed point is reached during the fast motion. The phase space trajectory is a straight line parallel to the x axis. Correspondingly, two branches of the line $P = 0$ are stable, and one branch is unstable.

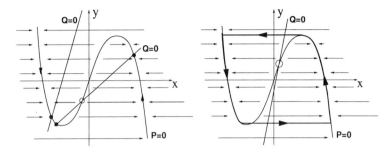

Figure 4: Phase plane of the FitzHugh-Nagumo model. Left: one or two stable fixed points; right: a limit cycle.

The slow motion happens on the line $P = 0$, and depends on the parameters a and b. If the intersection of $Q = 0$ and $P = 0$ happens on the stable branch, it gives a stable fixed point. If $Q = 0$ intersects with the unstable branch of $P = 0$, there are no stable fixed points in the system, and the only attractor is a stable limit cycle. This cycle consistsof two pieces of slow motion, and two pieces of fast motion.

An interesting state appears when $Q =$ and $P = 0$ intersect near the edge of the stable branch. In this situation relatively small perturbation of the stable state can lead to a large response – an impulse of finite length.

4   High-dimensional systems

The main feature that appears starting from dimension 3, is chaos. The corresponding attractor in the phase space is called *strange attractor*. Chaotic motion can be characterized as an irregular non-repeating process, sometimes it is called quasirandom. The main feature of chaos is its intrinsic instability: perturbations at the every point of a trajectory grow exponentially. Nevertheless, the motion as a whole remains an attractor, i.e. it attracts all neighbouring points. This is possible in 3 dimensions, where a volume shrinks while simultaneously it can disperse along some directions. Intuitevely, it is easy to understand, why instability causes chaos. Suppose we start at a point on the attractor. Soon or late the trajectory will return in the neighbourhood of the initial point (otherwise it were transient). After this happens there are two possibilities: if the initial part is stable, the trajectory will follow this part and the motion will be repeated; if the initial part is unstable, the trajectory will go away and will not repeat the initial part (see Fig. 5). Thus stability means

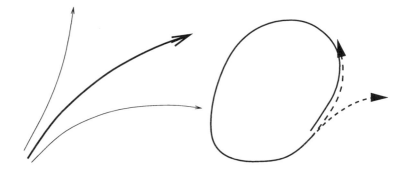

Figure 5: Left: instability of trajectories, right: instability implies non-regularity.

regularity (periodicity or, at least, quasiperiodicity), while instability implies chaos.

Example 4 (The Belousov-Zhabotinsky reaction) This is the chemical reaction that demonstrates typically periodic behavior. Under certain conditions, however, chaotic regimes can be observed, as is shown in Fig. 7.[7] When a parameter (the flow rate of the reagents) is changed, a complex sequence of transitions (bifurcations) between periodic and chaotic states happens. It is interesting that such a behavior can be modelled with relatively simple dynamical system:[8]

$$\frac{dx}{dt} = hx + y + ez \tag{7}$$

$$\frac{dy}{dt} = -x \tag{8}$$

$$\varepsilon\frac{dz}{dt} = f(x, z) \tag{9}$$

where $f(x, z) = 0$ defines a S-shaped surface and $\varepsilon \ll 1$. This system is similar to the FitzHugh-Nagumo system: the motions can be divided into fast (only the variable z changes) and slow (the motion on the surface $f(x, z) = 0$, called slow manifold). The slow motion looks like a growing spiral, when it reaches the border of the stable branch of the slow manifold a transition to another branch occurs. After that the trajectory returns to the vicinity of the unstable fixed point and the new spiralling begins (Fig.6). Note, that the attractor is bounded in the phase space: it consists of two pieces of the slow surface connected with fast jumps. One can see also, that there is a significant level of instability in the system: during growing spiralling the nearby points go away from each other. If this instability is not suppressed during the return process, one observes chaotic oscillations. When the parameter h is changed,

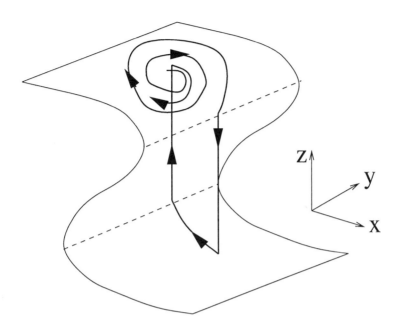

Figure 6: A sketch of the phase portrait of the model (7)-(9).

periodic regimes also appear, and in fact it is possible to reproduce in the model (7)-(9) the whole sequence of transitions observed experimentally.

5 Discrete-time systems

Dynamical systems with discrete time are very polpular because they allow one to describe complex states even when the dimension of the model is small. Roughly speaking, n-dimensional invertable (i.e. those that can be iterated both forward and backward) discrete-time systems have properties similar to $(n + 1)$-dimensional continuous-time models. If the invertibility condition is skipped, the dimension can be further reduced by 1. Thus, chaos can be observed already in 1-dimensional discrete-time models. The most popular is, of course, the logistic map

$$x(t + 1) = ax(t)(1 - x(t)) \tag{10}$$

which demonstrates, as the parameter a is increased, a transition from a stable fixed point to a periodic state, and further to chaos.

122

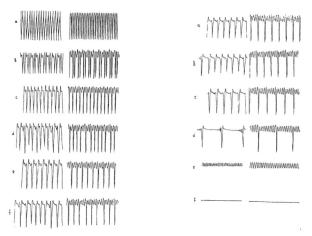

Figure 7: The experimental results from the Belousov-Zhabotinsky reaction (left column) and from the model (7)-(9) (right column).

6 Forced systems

There are two major types of external forcing: periodic and noisy. A periodic forcing adds one dimension to the phase space. Indeed, suppose that the r.h.s. of an ordinary differential equation (2) has a periodic time-dependence: $f(x, t) = f(x, t + T)$. Then Eq. (2) can be rewritten as a two-dimensional model, if a supplementary variable θ is introduced

$$\frac{dx}{dt} = f(x, \theta),$$
$$\frac{d\theta}{dt} = 1.$$

Many interesting effects occur if a system with periodic oscillations (limit cycle) is forced periodically; the period of the oscillations may be entrained by the external force.

Example 5 (The insulin regulation model[9]) In many experiments, concentrations of insulin and glucose oscillate with a characteristic period ≈ 2 hours. In special experiments non-diabetic men were subject to periodic glucose infusion. The observed profiles of insuline and glucose show characteristic "synchronization" phenomena, with the oscillations being entrained by the external periodic force (Fig. 8). A realistic but rather complicated model for the insuline/glucose feedback mechanism demonstrates similar behavior.

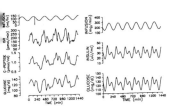 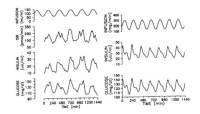

gure 8: Experimental (left column) and numerical (right column) results on the periodically rced insulin/glucose regulation.

While physical and chemical experiments are relatively noise-free, in bio-·gical systems irregular (noisy) evironment is inavoidable. Even small noise an lead to macroscopic dynamical effects, especially if the system is near the ;ability border. E.g., adding small noise to the FitzHugh-Nagumo equation ı the excitable case leads to the appearance of sustained oscillations (see ig. 9 [10]). Such a model has been used recently for modelling dysrythmias of ıe respiratory oscillator.[11]

Infinite-dimensional systems

[ere we briefly discuss two types of systems with infinite-dimensional phase ɔace. Diffrential-delay models describe dynamics where the evolution depends ot only on the state of the system at the same time, but on the state at some elayed time in the past. Models of such type have been proposed to describe ynamic deseases. While a healthy person has nearly constant number of blood ells, in some cases of leukaemia this number oscillates with a period of 70 days.

To model self-regulation of blood cells, Mackey and Glass suggested a ıodel with delay:

$$\frac{dx}{dt} = -x + \frac{bx(t-\tau)}{1+x^{10}(t-\tau)}$$

'o be able to solve this equation, we need to know the function $x(t)$ on the ıterval $[-\tau, 0]$, in this sense the system is infinite-dimensional. The attractor ;, however, low-dimensional. For small τ it is a stable fixed point, for larger elays – a limit cycle, and for sufficiently large values of τ a strange attractor ,ith chaotic oscillations is observed (Fig. 10).

Partial differential equations represent another importand class of systems ,ith infinite-dimensional phase space. Usually, one independent variable is ime, and the other is a spatial coordinate (however, one can consider other ituations, where, e.g., age distribution of the population is important).

124

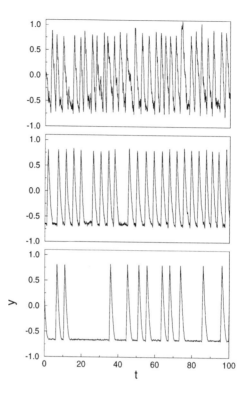

Figure 9: Regimes in the noisy FitzHugh-Nagumo for different noise amplitudes (noise grow from bottom to top, see Pikowski and Kurths (1997) [10] for details).

Example 6 (The Fisher-Kolmogorov-Petrovsky-Piskunov model.) is a parabolic nonlinear equation

$$\frac{\partial u}{\partial t} = au - u^3 + b + D\frac{\partial^2 u}{\partial x^2}. \tag{11}$$

The nonlinear function is taken from the example 1, we assume that the parameter a, b are such that the system has two stable states. Such a model describes a fire propagation. One stable state represents a combustable material, which, being heated can burn into another stable state. Diffusion is responsible for the propagation o the fire. If $D = 0$, an initially burning region does not influence its neighbours, whil for finite D the heat is spreading and a fire front can develop (Fig. 11). It can b shown, that Eq.(11) has solutions in the form of travelling with constant velocity

$$V = \sqrt{Da/2}(u_1 - 2u_2 + u_3)$$

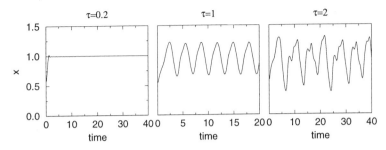

Figure 10: The dynamics of the Mackey-Glass systems for $b = 2$ in dependence on the delay time τ

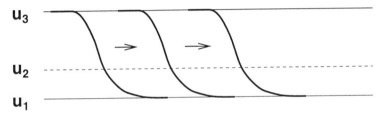

Figure 11: The front propagation described by model (11). The basin of the fixed point u_3 is larger than that of u_1, therefore the front moves to the right.

fronts, where u_i are the fixed points of the equation (2).

An interesting effect happens if one considers the distributed version of the FitzHugh-Nagumo system (5),(6):

$$\varepsilon\frac{\partial u}{\partial t} = -u^3 + u - v + D\frac{\partial^2 u}{\partial x^2} \tag{12}$$

$$\frac{dv}{dt} = u - bv + a \tag{13}$$

Let us choose the parameters a, b so that the system is in the excitable state: there is only one stable fixed point, but initial perturbation leads to a large deviation. Suppose that initially a small piece of the medium is excited. From this place two fronts of excitation similar to those in the Fisher-Kolmogorov-Petrovsky-Piskunov model begin to propagate (this is fast motion where only u changes). Due to slow evolution of the variables u, v the excited state disappears after some time, and the system evolves back to the stable fixed point. A pulse of excitation is then formed consisting of two parts: the excitation of the variable u, and the tail where the variable v returns to the fixed point and the system is not excitable. When interpreting the pulse as a fire, the variable

126

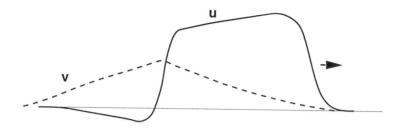

Figure 12: The isolated pulse in the FitzHugh-Nagumo model with diffusion.

v corresponds to the grass which grows slowly, and until new grass is grow
a new fire is impossible. In two spatial dimensions beuteful patterns of spir
waves appear.

8 Numerical modelling

There are no large difficulties in modelling dynamical systems. Computer im
plementation of discrete-time systems (maps) is straightforward. There ar
many reliable methods to solve ordinary differential equations, the most popu
lar are the Runge-Kutta methods, and predictor-corrector methods (see Pres
et al. (1989) [12]). One can meet a difficulty if the system has different tim
scales, as the FitsHugh-Nagumo model (5),(6) has. In this case one shoul
use methods of integration of so-called stiff systems. There are also integrate
computer packeges designed for investigation of dynamical systems (interactiv
bifurcation analysis etc.). [13,14]

Modelling noisy dynamical systems is slightly nontrivial. Consider a
equation

$$\frac{dx}{dt} = f(x) + D\xi(t)$$

where $\xi(t)$ is a noisy term, which is assumed to be Gaussian δ-correlated wit
zero mean. Let us write an approximate solution of this equation on th
interval $[t, t + \Delta t]$:

$$x(t + \Delta t) \approx x(t) + f(x(t))\Delta t + D \int_t^{t+\Delta t} \xi(t)\, dt$$

The last term, as a sum of Gaussian random variables, has Gaussian distribu
tion and the variance $D^2\Delta t$. Thus, the approximate solution can be writte
as

$$x(t + \Delta t) \approx x(t) + f(x(t))\Delta t + D\sqrt{\Delta t}\eta_t$$

here η_t are independent Gaussian random variables with unit variance. Higher-order methods for integration of noisy systems also exist.

Numerical modelling of partial differential equations is a subject of its own, see Press *et al.* (1989) [12] for introduction to it.

Some remarks are in order on the modelling of differential-delay equations. Consider the equation

$$\frac{dx}{dt} = -x + f(x(t-1))$$

The best way to discretize this equation is to use an implicit difference method, namely to calculate the r.h.s. at the future time step:

$$\frac{x(t+\Delta t) - x(t)}{\Delta t} = -x(t+\Delta t) + f(x(t-1+\Delta t))$$

which, fortunately, can be rewritten as an explicit scheme:

$$x(t+\Delta t) = \frac{x(t) + \Delta t f(x(t-1+\Delta t))}{1+\Delta t}$$

Inverse problems and data analysis

Proving that a real experimental system can be described by a dynamical model is a challenge. This topic is a subject of *nonlinear data analysis*. The main idea is as follows: suppose that we have a process $x(t)$, the task is to reconstruct a dynamical system that produces $x(t)$. Note first, that what is observed in experiments is an attractor, therefore one can hardly reconstruct transient behavior if the experiment is not specially planned. An attractor reconstruction has sense if the regime observed is nontrivial. Indeed, one does not need to reconstruct a fixed point or a limit cycle. A chaotic motion is nontrivial, and the main question is how to distinguish it from noise. First of all, one has to do with a high-dimensional phase space, while in experiments usually one or few variables is availible. This difficulty can be overcome with the *embedding* (Takens') method: one considers time-delayed variables $x(t-\tau)$, $x(t-2\tau)$ as independent variables in the phase space. Already drawing such a phase space may help to recognize a dynamical structure. More elaborated methods are currently a subject of intensive studies. [15]

1. L. Glass and M.C. Makcey, *From Clocks to Chaos. Thee Rhytms of Life* (Princeton Univ. Press, Princeton 1988).
2. J. Guckenheimer and P. Holmes, *Nonlinear Oscillations, Dynamical systems, and Bifurcations of Vector Fields* (Springer, N.Y. 1986).

3. S. Wiggins, *Introduction to Applied Nonlinear Dynamical System* (Springer, Berlin 1989).

4. P. Glendinning, *Stability, instability and chaos* (Cambridge Universit Press, Cambridge 1994).

5. Yu. Kuznetsov, *Elements of Applied Bifurcation Theory* (Springer Ap plied Mathematical Sciences Series, Vol. 112, 1995).

6. M. Eisen, *Mathematical methods and models in the biological science* (Prentice Hall, Englewood Cliffs 1988).

7. J.L. Hudson, M. Hart, and D. Marinko *An experimental study of multipl peak periodic and nonperiodic oscillations in the Belousov-Zhabotinsk reaction*, J.Chem.Phys., **71**, 1601–1606 (1979).

8. A.S. Pikovsky, *A dynamical model for periodic and chaotic oscillation in the Belousov-Zhabitinsky reaction*, Phys. Lett. A **85**, 13–16 (1981).

9. J. Sturis, C. Knudsen, N.M. O'Meara, J.S. Thomsen, E. Mosekilde E. VanCauter, and K.S. Polonsky, *Phase-locking regions in a forced mode of slow insulin and glucose oscillations*, CHAOS **5**, 193–199 (1995).

10. A. Pikovsky and J. Kurths *Coherence resonance in a noise-driven ex citable system*, Phys. Rev. Lett. **78**, 775-778 (1997).

11. D. Paydarfar and D.M. Burkel, *Dysrythmias of the respiratory oscillator* CHAOS **5**, 18–29 (1995).

12. W.H. Press, B.P. Flannery, S.A. Teukolsky, and W.T. Vetterling, *Nu merical Recipes*, (Cambridge University Press, Cambridge 1989).

13. M.A. Taylor and I.G. Kevrekidis, *Interactive AUTO: A graphical inter face for AUTO 86*, Technical report, department of chemical engineering (Univ. Press, Princeton NJ 1990).

14. U. Feudel and W. Jansen, *Candys/qa-a software system for qualitative analysis of nonlinear dynamical systems*, J. Bifur. Chaos **2**, 773–79 (1992).

15. P. Grassberger, T. Schreiber, and C. Schaffrath, *Nonlinear time sequence analysis*, Int. J. of Bifurcation and Chaos **1**, 521–547 (1991).

NONLINEAR EXCITATIONS AND ENERGY LOCALIZATION. APPLICATIONS TO MOLECULAR BIOLOGY

MICHEL PEYRARD

Laboratoire de Physique de l'Ecole Normale Supérieure de Lyon
CNRS URA 1325, 46 allée d'Italie, 69007 Lyon, France.

Introduction

Modern molecular biology has unraveled the extreme complexity of biological processes at the molecular level. Current techniques allow biologists to observe thinner and thinner details of these processes and this leads naturally to the idea of *biological diversity*. One key concept in biology is *specificity*. In contrast the aim of science is to look for *general laws*. In the word of Poincaré "Il n'y a de science que du général".[1]

A simple example can illustrate how a problem can be considered from very different view points. Let us assume that our object of study is a car engine, a system much simpler than the simplest biological system. One can look at it with the eyes of the engineer and describe cylinders, carburettor, valves, and, as our knowledge expands, we would discover thinner details such as the small screw that regulates the gas-air mixture. Then, by turning this screw we would be able to make it run better or worse and this would give us the feeling that we begin to understand and control this complicated machinery. But if we are given another engine, from another brand, we may have to start over. It may not have a screw for mixture tuning, or it may have two. As a physicist, our view would be very different. We would look for basic principles such as the thermodynamic cycle of the gas-air mixture that expands and pushes pistons, or the way a crank can turn the oscillatory motion of the pistons into the rotation of the driving shaft. This knowledge is valid for any engine, but, if his car breaks down the physicist may not be able to repair it! However from his understanding he may be able to suggest that, if we add a turbo-compressor to increase the admission pressure, we get a much more powerful engine.

Thus both approaches are complementary. Currently biologists are essentially applying the engineer's approach to nature and they have learned how to control some of its processes to an impressive accuracy. But the underlying fundamental mechanisms are still to be discovered. My aim in these lectures is to introduce concepts and tools that could help us to exhibit some general principles in molecular biology. These ideas may also tell us which level of description is necessary (and sufficient) to study a given phenomenon. For instance current molecular dynamics studies of DNA can hardly extend beyond 10 or 20 base-pairs and a few pico-seconds. This is not enough to model transcription, but perhaps we do not need all the details provided by molecular dynamics to understand the main features of transcription.

Looking for general laws is what physics is all about. At the deepest level, it has developed theories such as mechanics, electromagnetism, quantum mechanics which are of general validity. These theories have been applied successfully to complex systems such as solids which contain a huge number of individual particles, and they can predict accurately for instance the electron energy bands. This knowledge has then been exploited to build devices like transistors, lasers, etc. Because biological molecules are made of the same particles as the materials that the physicist studies, *they are physical systems* which obey the same general laws. Therefore one may think that the problem is solved: we know the laws, and we even know how to apply them to a system made of many particles such as a crystal or even a disordered system such as a glass so that it is sufficient to do the same for biological molecules to understand how biological systems operate. This is however not exactly true. These general laws of physics are defined at such an elementary level that there is a long way between the description at the particle level to the understanding of a collective system. *It is here that biophysics differs from material science.* The fundamental laws are the same but the context is different. Solid state physics has developed many-body theory but the equivalent approach for biological molecules is still under investigation. Let us compare for instance the dynamics of a crystal or a biological molecule. The crystal is a very ordered, rigid, object. A disturbance at one end of its lattice will propagate along and it is well described by small amplitude linear waves, or phonons. On the contrary any segment of the biological molecule is different from its neighbors. Moreover the molecule is a very deformable object which can undergo large conformational changes. Therefore the concept of extended linear waves or phonons is likely to be irrelevant for biological molecules. Nevertheless it is often used because it became so familiar in condensed matter physics.

In the last three decades nonlinear science has introduced new concepts which may be more suitable for the description of the physical processes governing the basic phenomena of life at the molecular level. The aim of these lectures is to introduce some of these new tools. A warning is however necessary here. It is not because we have new concepts that look very powerful that we should pretend that we "understand" how living organisms operate. But the ideas that emerged from nonlinear science could provide a better basis for this understanding than the more familiar method of solid state physics that made their way into biophysics.

The first part of the lectures will introduce the concept of *soliton* as a paradigm that is applicable to many physical phenomena. We shall also examine how it differs from the phonons and why it could provide a better basis for the description of some biological phenomena, and point out some of its limitations. The study of proton transport in hydrated proteins will illustrate one application of the soliton concept to investigate collective phenomena in a biophysical system. Then we shall examine how the solitons can be formed spontaneously in a system. This will lead us to the idea of nonlinear energy localization which can be applied to analyze DNA thermal denaturation and transcription.

2 Solitons: concept and basic properties.

The response of *most* of the physical systems to combined excitations is *not a simple superposition* of their response to individual stimuli. Mathematically, this property means that almost all physical and biological systems are *nonlinear*. Although this has been recognized for a long time, many theoretical models in physics and biology are still relying on a *linear description*, corrected as much as possible for nonlinearities which are treated as small perturbations. It is now well known that such an approach can be *qualitatively wrong*. Although linear response theory can give useful approximate results to analyze some experimental data, the linear approach can sometimes miss completely some essential behaviors of the system. This is particularly true for biological systems in which the nonlinear effects are often the *dominant ones.* For instance many biological reactions would not occur without large conformational changes which cannot be described, even approximately, as a superposition of the normal modes of the linear theory.

The *intrinsic* treatment of nonlinearities in mathematical models, and later in physical systems, has renewed completely our view of some phenomena. In particular, two concepts have emerged: chaos and solitons. Chaos, which can appear in a nonlinear system with a small number of degrees of freedom, has been extensively discussed, even in non scientific journals, because it challenges our belief that a system described by a small number of deterministic equations should be easy to understand and predictable. But the concept of *soliton*, relevant for systems with many degrees of freedom which can cooperate to form a coherent nonlinear excitation, is also very important, particularly in the context of biology. It is tempting to say that these self-organized dynamical structures, which can emerge in a complex system from a large variety of excitations, provide a paradigm for similar behavior in biological systems. Although this analogy must certainly be taken with caution, the soliton provides nevertheless a tool which should not be ignored in the studies of biomolecules because it has no equivalent in the world of linear excitations.

2.1 The concept of soliton.

The first observation of a soliton was made by J. Scott-Russel in 1834. He was looking at a boat moving on a shallow canal and noticed that, when the boat suddenly stopped, the wave that it was pushing at its prow took its own life: "it accumulated round the prow of the vessel in a state of violent agitation, then suddenly leaving it behind, rolled forward with great velocity, assuming the form of a large solitary elevation, a rounded, smooth and well defined heap of water which continued its course along the channel apparently without change of form or diminution of speed". J. Scott-Russel followed the wave along the canal for several kms. He was so impressed by this "great solitary wave", as he called it, that he spent 10 years of his life to study experimentally the properties of this "singular and beautiful phenomenon". Whatever the actual experiment which is made to show a soliton, it always impresses the observer because the properties of the soliton are against our intuition built on the observation of small amplitude linear waves that spread over as they propagate. On the contrary the soliton is extremely robust. The analytical explanation of Scott-Russell's observations was only provided much later with the works of Boussinesq in 1972 and Korteweg–de Vries in 1895. The phenomenon was then forgotten until 1965 when N. Zabusky and M.D. Kruskal coined the word "soliton" to describe this solitary wave which has particle-like properties. They also provided a fundamental understanding of the origin of the soliton and initiated the development of mathematical methods which started a "soliton industry" in applied mathematics.

A simple example: the pendulum chain.

A simple example is provided by the chain of pendula coupled by a torsional spring shown in fig 1. This regular lattice can be viewed as a model crystal in which molecules can rotate or as a representation of one of the strands of DNA with its bases attached on the side.

If the first pendulum is displaced by a small angle θ_1, this disturbance propagates as a small amplitude linear wave from one pendulum to the next through the torsional coupling. As it moves along the chain, the small amplitude localized perturbation spreads over a larger and larger domain due to dispersive effects. In order to describe this phenomenon quantitatively, we need the equations of motions of the pendula. The chain involves only simple mechanics so that it is easy to write its equations of motion. Its energy is the

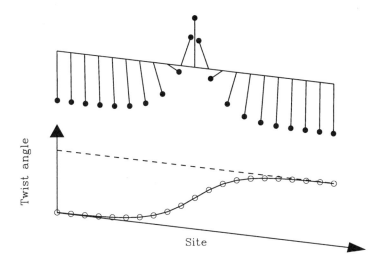

Figure 1: Soliton in a chain of pendula coupled by a torsional spring. The soliton is a 2 π rotation. The upper part shows the aspect of the pendulum chain and the lower figure shows the rotation angle of the pendula as a function of their position along the chain.

sum of the rotational kinetic energy, the elastic energy of the torsional spring connecting two pendula, and the gravitational potential energy. Denoting by θ_n the angle of deviation of pendulum n from its vertical equilibrium position, by m its mass and l the distance between the rotation axis and its center of mass, the expression of the energy is then

$$H = \sum_n \frac{1}{2} I \left(\frac{d\theta_n}{dt} \right)^2 + \frac{1}{2} C (\theta_n - \theta_{n-1})^2 + mgl(1 - \cos\theta_n) , \qquad (1)$$

where I is the moment of inertia of a pendulum around the axis, and C is the torsional coupling constant of the springs. The equations of motion which can be deduced from this hamiltonian are

$$\frac{d^2\theta_n}{dt^2} - c_0^2(\theta_{n+1} + \theta_{n-1} - 2\theta_n) + \omega_0^2 \sin\theta_n = 0 , \qquad (2)$$

with $c_0^2 = C/I$ and $\omega_0^2 = mgl/I$. This simple looking set of differential equations has *no known analytical solution*. We encounter here a frequent situation in nonlinear science because nonlinear differential equations (or partial differential equations) are generally very difficult to solve. The common attitude in this case is to *linearize* the equation, i.e. to replace $\sin\theta_n$ by its small amplitude expansion $\sin\theta_n \approx \theta_n$. We get

$$\frac{d^2\theta_n}{dt^2} - c_0^2(\theta_{n+1} + \theta_{n-1} - 2\theta_n) + \omega_0^2\theta_n = 0 \ . \tag{3}$$

Now the set of equations is easy to solve. It has wavelike solutions

$$\theta_n = \alpha\exp[i(\omega t - qn)] \ , \tag{4}$$

where α is the amplitude of the wave, ω its frequency and q its wavevector. This is a solution if ω and q are related by the relation

$$\omega^2 = \omega_0^2 + 2c_0^2(1 - \cos q) \tag{5}$$

as one can easily check by putting this solution in Eq. (3). These linear waves are the "phonons" of the solid state physicist. This approach is justified if we apply only small excitations to the system, or if the chain is so rigid that even fairly large excitations generate only small displacements. This is the case of most of the crystals. But biological molecules are very deformable objects which often exhibit large conformational changes. To mimic such a behavior, one can imagine to rotate the first pendulum by a full turn. *Then we get the soliton* which is much more spectacular to observe than the linear waves. The 2π rotation propagates along the whole pendulum chain and even reflects at a free end to come back unchanged. With such a simple mechanical device, it is easy to check the exceptional properties of the solitons. Launching a soliton and keeping on agitating the first pendulum we can test the ability of the soliton to propagate over a sea of linear waves. It is even more remarkable to observe that the solitons have particle-like properties, and this explains why their name ends like the name of many elementary particles (electron, proton, neutron etc). If one static soliton is created in the middle of the chain and then a second one is sent, the collision looks to the observer exactly similar to a shock between elastic marbles.

This experiment demonstrates that linearizing the equations of motion to get Eq. (3) *had lost essential physics.* The linearized equations have no localized solutions and they have no chance to describe the soliton even approximately because $\theta_n = 2\pi$ is not a small angle!

Although we do not know how to solve exactly Eqs. (2), there is however another possibility to solve them approximately, while preserving their full nonlinearity. If the coupling between the pendula is strong enough, adjacent pendula have similar motions and the discrete set of variables $\theta_n(t)$ can be replaced by a single function of two variables, $\theta(x,t)$ such that $\theta_n(t) = \theta(x =$

x, t). A Taylor expansion of $\theta(n+1, t)$ and $\theta(n-1, t)$ around $\theta(n, t)$ turns the discrete set of equations (2) into the partial differential equation

$$\frac{\partial^2 \theta(x, t)}{\partial t^2} - c_0^2 \frac{\partial^2 \theta(x, t)}{\partial x^2} + \omega_0^2 \sin \theta = 0 . \tag{6}$$

This equation is called the "sine-Gordon" equation and it has been extensively studied in soliton theory because it has exceptional mathematical properties. In particular, it has a soliton solution

$$\theta(x, t) = 4 \tan^{-1} \exp \left[\pm \frac{\omega_0}{c_0} \frac{x - vt}{\sqrt{1 - v^2/c_0^2}} \right] , \tag{7}$$

which is plotted in Fig. 1. This solution provides a very accurate description of the 2π-torsion propagating as a soliton along the pendulum chain.

The example of the pendulum chain shows clearly that the soliton is a localized packet of energy. Away from its center, the pendula are in their rest position which can be chosen as the reference energy level. But, inside the moving soliton, the three types on energy terms appearing in the hamiltonian (1) contribute to raise the energy: the rotating pendula have kinetic energy, the torsion is associated to an elastic energy in the springs and the pendula involved in the soliton have extra gravitational energy. The soliton solution introduced into the hamiltonian shows that the energy falls off exponentially away from the soliton center.

The pendulum chain has a double interest. First it provides an experimental device which convincingly demonstrates the properties of the soliton. But it also illustrates a new approach to a theoretical description of a physical problem which can be very fruitful. When one is confronted to a set of complicated equations describing a physical or biological phenomenon, *it is important to remember that linearizing the equations to get an approximate solution is not always a good answer.* The continuum limit approximation that we have discussed above is one alternative. Another one which is very fruitful is the multiple scale expansion [2,3] which will be discussed in section 4.

General properties of solitons.

The experimental observations of solitons on a pendulum chain have exhibited some of their properties. More generally, solitons are *solitary waves*, i.e. waves localized in space, which have very special properties:

- they propagate at constant speed without changing their shape,

136

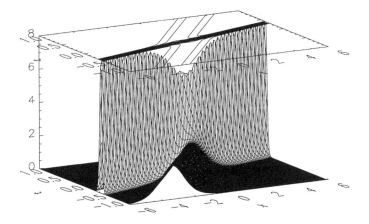

Figure 2: Collision of two solitons, solutions of the Korteweg–de Vries equation describing shallow water waves. The vertical scale is highly magnified. Real shallow water solitons have an amplitude which is small with respect to their width.

- they are extremely stable to perturbations, and in particular to collisions with small amplitude linear waves,

- they are even stable with respect to collisions with other solitons as illustrated in Fig. 2. In such a collision they pass through each other and recover their speed and shape after the interaction. This is surprising because one might think that the strong nonlinearity would break up the pulses during the interaction which is indeed not a simple superposition of the two waves. This can be observed directly on Fig. 2 because at the collision point, the amplitude of the wave is *not* the sum of the amplitudes of the two incoming waves. Moreover, after the collision the trajectories of the two excitations have been shifted with respect to trajectories without the collision. It is however the "soliton miracle" that finally, the outcome of the collision of two solitons is a simple phase shift of each excitation.

Solitons are not restricted to hydrodynamics or the pendulum chain. As discussed below, they can can appear in a large variety of systems.

2.2 Conditions to have solitons.

The equation established by Korteweg and de Vries (now known as KdV equation) to describe the soliton observed by Scott-Russel at the surface of a canal illustrates clearly the conditions that a system must fulfill in order to sustain solitons. Calling $u(x,t)$ the height of the wave above the free surface at equilibrium, the KdV equation is

$$\frac{\partial u}{\partial t} + \frac{3}{2h}u\frac{\partial u}{\partial x} + \frac{h^2}{6}\frac{\partial^3 u}{\partial x^3} = 0 \,, \tag{8}$$

where h is the depth of the water in the canal. This equation is written in a frame moving at the speed of long-wave linear disturbances of the surface.

Let us suppose for a moment that the second term which contains the product $u(\partial u/\partial x)$ can be ignored. This term is the nonlinear term of the KdV equation, and, without it, the linearized KdV equation,

$$\frac{\partial u}{\partial t} + \frac{h^2}{6}\frac{\partial^3 u}{\partial x^3} = 0 \,, \tag{9}$$

has plane wave solutions of the form

$$u = u_0 \exp[i(qx - \omega t)] \,. \tag{10}$$

Putting such a solution into equation (9), one gets the dispersion relation of the wave $\omega = -h^2 q^3/6$. Any initial disturbance of the surface can be decomposed into its Fourier components (10) and the velocity of each component in the moving frame is is $v(q) = \omega/q = -h^2 q^2/6$. Threfore each component has its own velocity, or, in other words, the medium is *dispersive*. Consequently, without the nonlinear term, a localized perturbation of the surface would tend to spread over as it propagates as shown schematically on fig 3 and the KdV equation would not support solitons. But the effect of the nonlinear term is exactly opposite. If we now forget temporarily the dispersive term, but keep the nonlinear term, the KdV equation reduces to

$$\frac{\partial u}{\partial t} + \frac{3}{2h}u\frac{\partial u}{\partial x} = 0 \,. \tag{11}$$

To understand the behavior of this equation, it is convenient to consider an even simpler equation

$$\frac{\partial u}{\partial t} + v_0\frac{\partial u}{\partial x} = 0 \tag{12}$$

which has traveling wave solutions $u = f(x - v_0 t)$ propagating at speed v_0. This suggests that the solutions of the truncated KdV equation (11) can be viewed

138

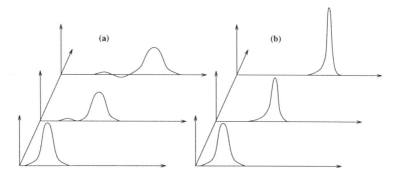

Figure 3: Schematic picture of the time evolution of a localized pulse in a system showing dispersive effects (a) or nonlinear effects (b).

as waves for which the speed $v_0 = 3u/2h$ is proportional to the amplitude u As shown in Fig. 3b, the front of such a wave tends to steepen because the crest of the wave moves faster than the bottom. If nonlinearity were acting alone, this steepening would lead to the formation of a shock wave and then to a breakup of the wave.

But the KdV equation contains *both dispersion and nonlinearity. Their balance is responsible for the existence of the solitons.* While nonlinearity tends to localize energy, dispersion tends to spread it over and the "miracle of the soliton" is that this balance is stable: if a wave is too broad, its Fourier spectrum is very narrow and dispersion plays only a small role while nonlinearity tends to win and make the wave steeper until the dispersion, which grows as the wave gets more localized, balances the nonlinearity. Similarly, for a wave which is initially too narrow, the huge dispersion causes it to become wider until the dispersion is low enough to be balanced by the nonlinear effects. Therefore, what may appear as a first glance as a fragile equilibrium is in fact a mechanism which guarantees the robustness of the soliton. For surface waves the KdV equation shows that the dispersion and nonlinearity are governed by the depth h of the water. The nonlinear term, proportional to $1/h$ increases for shallow water, while the dispersion, proportional to h^2 decreases. While small fluctuations of h do not perturb the soliton, the equilibrium between the two can however be achieved only is h is roughly constant. This is not true for waves approaching a beach because h decreases continuously in the frame of the wave and nonlinearity finally wins: the wave breaks and rolls over, a phenomenon well known of people practicing surf riding.

This discussion on stability shows that the balance between dispersion and nonlinearity can not only explain the *existence* of solitons but also their *for mation* from a wide range of localized initial conditions. This is a very general property which can also be found in the pendulum chain for instance. Here the situation is however slightly more subtle because the same term $\omega_0^2 \sin \theta$ of the sine-Gordon equation contains dispersion and nonlinearity. Its nonlinear character comes from the sinusoidal function. Its role in the dispersion can by observed by linearizing the sine-Gordon equation which becomes

$$\frac{\partial^2 \theta(x,t)}{\partial t^2} - c_0^2 \frac{\partial^2 \theta(x,t)}{\partial x^2} + \omega_0^2 \theta = 0 . \tag{13}$$

A plane wave $\theta = \theta_0 \exp[i(qx - \omega t)]$ has the dispersion relation $\omega^2 = \omega_0^2 + c_0^2 q^2$, so that, for $\omega_0^2 \neq 0$, ω/q is not a constant and the medium is dispersive for the wave.

Besides the balance between dispersion and nonlinearity, equations having exact soliton solutions have many nice mathematical properties which have delighted many mathematicians. [2,4] They can have multi-soliton solutions and there exists a nonlinear analogous of the Fourier transform for linear equations which allows the decomposition of a given signal into its soliton content. This "inverse scattering transform" has been used for instance to analyze experimental observations of solitary waves created by the tide in the Andaman sea near the coast of Thailand. [5] These quasi-solitons can be observed from satellites and they propagate over hundreds of kilometers in a shallow sea. This example provides an impressive illustration of the soliton concept, and it also shows its power. Although the bottom of the Andaman sea is not flat, although the coast is far from being as straight as the side of the Scott-Russel's canal, the KdV equation provides a very good framework to analyze quantitatively the behavior of these large, extremely long lived, waves that any linear theory simply cannot explain.

2.3 The different classes of solitons.

Solitons can be divided into two main classes, topological and non-topological solitons.

The 2π-rotations of the pendulum chain provide an example of *topological* solitons. In order to understand this terminology, it is convenient to plot the gravity potential acting on the pendula, versus θ and the position x of a pendulum (Fig. 4). Since all the angles $\theta = 0, 2\pi, 4\pi, \ldots$ correspond to the minimal energy of the pendula, the physical system has *degenerate energy minima*. The topological soliton is an excitation which interpolates between these minima. It can exist at rest and is extremely stable because, in an infinite system, it can only be destroyed by moving a semi-infinite segment

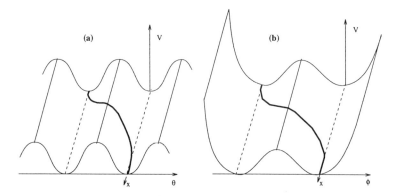

Figure 4: Representation of topological solitons showing the shape of the on-site potential as a two-dimensional surface depending on the soliton variable and on position. The heavy line shows the trajectory of the soliton variable on this potential energy surface for a soliton solution. (a) The sinusoidal potential of the pendulum chain. (b) The double-well potential of the ϕ^4 model.

of the system above a potential maximum. This would require an infinite energy. The system can have solitons (going from one minimum θ_0 to another minimum θ_0', but also anti-solitons going from θ_0' to θ_0. The topological soliton could only be destroyed by a collision between a soliton and an antisoliton. In an integrable system having exact soliton solutions, solitons and anti-solitons simply pass through each other with a phase shift, as solitons do, but in a real system like the pendulum chain which has some dissipation of energy, the soliton–antisoliton equation may destroy the nonlinear excitations.

The example of the pendulum chain shows that the condition for the existence of the topological soliton is simply the existence of degenerate minima of the gravitational potential $V(\theta) = mgl(1 - \cos\theta)$. Two minima are sufficient for the existence of the topological soliton, as illustrated in fig 4b for another typical model, the "ϕ^4" model where the potential $V(\phi) = V_0(1 - \phi^2)^2$ replaces the gravitational energy (in this case, it is customary to call ϕ the soliton variable, instead of θ). Another remakable feature of the topological solitons is that they are Lorentz invariant with respect to the maximum group velocity c_0 of the linear waves in the system. This is reflected in the factor $\gamma = 1/\sqrt{1 - v^2/c_0^2}$ which appears in the solution of the sine-Gordon equation (7). Consequently the soliton cannot propagate faster than c_0 and fast solitons show a "relativistic" contraction which is easily visible in the experiments performed with the pendulum chain.

Contrary to the topological solitons of the pendulum chain, for the KdV

(a)

(b)

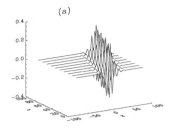
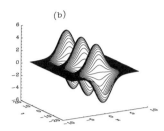

Figure 5: (a) Propagation of a soliton solution of the NLS equation. (b) Time evolution of a breather, solution of the sine-Gordon equation.

solitons of water waves, the state of the system is the same on both sides of the soliton. Such *non topological* solitons are dynamical entities. They cannot exist at rest. The amplitude of the KdV solitons is related to their velocity v and to the maximum speed c_0 of linear waves by $u_{max} = \sqrt{v^2 - c_0^2}$ which shows that these solitons are necessarily supersonic. Moreover their equation of motion is galilean invariant instead of Lorentz invariant for the topological solitons.

The two examples that we have discussed up to now are *permanent profile solitons*. This is not always so and some solitons have an internal dynamics. An important example is provided by the solitons of the "Nonlinear Schrödinger" (NLS) equation

$$i\frac{\partial A}{\partial t} + P\frac{\partial^2 A}{\partial x^2} + Q|A|^2 A = 0 , \qquad (14)$$

where P and Q are constant coefficients. This equation describes the time evolution of the complex amplitude A of a weakly nonlinear wave. If $PQ > 0$, it has soliton solution which are localized wavepackets as shown in Fig. 5a. The relative motion of the envelope and carrier wave is responsible for the internal dynamics of the NLS soliton. Another example of a soliton with internal dynamics is provided by the "breather" of the sine-Gordon equation. It is a large amplitude oscillation in the bottom of one of the potential valleys of Fig. 4a. Its time evolution is shown in Fig. 5b.

These examples show that the world of solitons includes various types of localized nonlinear excitations. Although they may look very different, they have in common their exceptional stability and particle-like properties which distinguish them from the linear waves that we are used to consider. In fact, real systems do not carry exact soliton solutions in the strict mathematical sense (which implies an infinite life-time and an infinity of conservation laws) but "quasi-solitons" which have most of the features of true solitons. In particular, although they do not have an infinite lifetime, quasi-solitons are generally so long-lived that their effect on the properties of the system are almost the same as those of true solitons. This is why physicists often use the word soliton in a loosly way which does not agree with mathematical rigor. Following this habit, we use henceforth the denomination soliton for quasi-solitons in real systems.

2.4 Solitons are everywhere!

This title has been used once in a scientific journal to show how the concept of soliton is ubiquitous and powerful. Examples of solitons can be found at all scales (from the hydrodynamic solitons of the Andaman sea to domain walls which are only a few crystal cell wide in ferroelectric materials) and in various domains of physics and chemistry. Let us consider a few typical examples where the soliton paradigm has been involved in the explanation and prediction of physical phenomena.

The sine-Gordon equation (6) does not only describe torsions in a pendulum chain. It provides also a very accurate description of "fluxons", i.e. quanta of magnetic flux, in Josephson transmission lines. A Josephson junction is a sandwich made of a thin dielectric film between two supraconductors (Fig. 6a). The potential difference V across the dielectric is related to the phase difference θ between quantum states of the Cooper pairs in the two supraconductors by

$$\frac{d\theta}{dt} = \frac{2e}{\hbar} V \ . \tag{15}$$

Moreover the tunneling of the Cooper pairs across the dielectric layer generates a current density j of the form $j = j_0 \sin \theta$, where j_0 is a current characteristic of a junction. Taking into account the inductance and the capacitance of the junction, one finally gets a sine-Gordon equation for $\theta(x, t)$, forced by a right hand side term which is associated to the current imposed through the junction by an electrical generator (bias current). The sine-Gordon solitons describe quanta of magnetic flux, expelled from the supraconductors, that travel back and forth along the junction. Their presence, and the validity of the soliton description, can be easily checked by the special shape of the current voltage characteristics of the junction as well as by the microwave emission which is associated to their reflections at the two ends of the junction.

While the solitons of the Josephson junctions are just beginning to be used in practical devices, optical solitons are moving fast to a multimillion dollar industry. They can be generated in optical fibers in which the optical index n

depends on the amplitude of the electric field through high order terms of the dielectric tensor

$$n(E) = n_0 + \chi_3 |E|^2 . \tag{16}$$

Introducing this expression into the Maxwell equations describing the electric field in the fiber, one gets a NLS equation for the amplitude A of the electric field of a plane wave of frequency ω_0 and wavevector k_0.

$$i\frac{\partial A}{\partial x} - \frac{1}{2}\left(\frac{\partial}{\partial \omega}\frac{1}{v_g}\right)\frac{\partial^2 A}{\partial t^2} + \frac{\chi_3 \omega_0^2}{2k_0 c^2}|A|^2 A = 0 \tag{17}$$

The coefficient of the second term is the dispersion of the group velocity v_g in the fiber while the nonlinear contribution is in the third term. In some frequency range, this equation has soliton solutions which can be used to carry information on extremely long distances at a very high rate. Moreover the exceptional robustness of the soliton to external perturbations can be used very efficiently to increase drastically the signal to noise ratio by the introduction of filters at regular intervals in the optical path so that the wavelength which they select changes slightly from one filter to the next. A small amplitude signal which is able to pass through one filter does not match the wavelength of the next one and is therefore stopped. On the contrary, the soliton is able to adjust itself to the perturbation and to propagate through the filters. It is remarkable that optical solitons, which are expected to be so widely used in applications in the near future, have been suggested from theoretical considerations. Nonlinear theory is still at the forefront of the industry for the development of new devices.

As mentioned above, solitons are not restricted to the macroscopic world. They show-up clearly in the properties of one dimensional magnetic materials. In compounds such as TMMC ($(CH_3)_4NMnCl_3$), the crystal structure is such that spins interact strongly along one axis of the crystal and very weakly along the other axes. These spin chains are qualitatively similar to the pendulum chain described above, and a torsion of the spin lattice can propagate as a soliton in the crystal. Figure 6b shows an example of such a magnetic soliton. These solitons are created thermally and their dynamics can be studied very accurately using magnetic resonance or neutron diffraction. In order to analyze the experimental results of such experiments, it is necessary to study the statistical mechanics of the magnetic chain. The soliton concept is again precious because the solitons can be treated as quasi-particles and the results can be obtained by investigating a "soliton gas".

Another example which has attracted a lot of attention is the case of conducting polymers. The simplest example is polyacetylene and, as shown

(a)

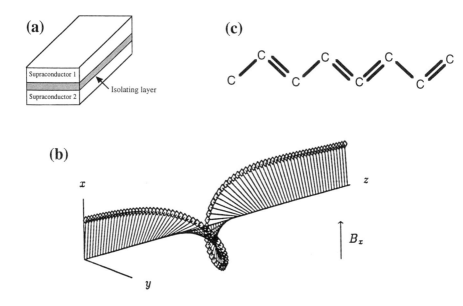

Supraconductor 1

Isolating layer

Supraconductor 2

(c)

(b)

x

z

B_x

y

Figure 6: (a) A long Josephson junction. (b) Magnetic soliton in a ferromagnetic spin chain. (c) Formula of polyacetylene showing the possibility to have a topological defect.

in Fig. 6c, its structure can exhibit solitons. The polymer chain has two degenerate states in which the positions of simple bonds and double bonds are exchanged. A defect interpolating between these two states can be treated as a topological soliton. It carries an electric charge and, due to the high mobility of the soliton, it can be responsible for a high electrical polymer conductivity. Following the original analysis,[7] the model has been refined and it has been shown that the most probable defects are not the topological solitons but breathers (or polarons) which are simply another class of nonlinear excitations. The breathers generate electronic states in the gap between the valence and conduction energy bands of the electrons. These states have been observed in infra-red experiments, showing the validity of the nonlinear treatment of conducting polymers.

There are many other examples where the soliton concept has been used to analyze experimental results in macroscopic physics (for instance nonlinear electrical lines) or solid state physics (dislocations,[8] domain walls in ferro-electrics,[9] charge transport in hydrogen bonded chains,[10] etc). Some examples in biology are discussed below but, in this domain, a lot of work has still to be done to derive models which are sufficiently accurate while staying tractable, and which can be confronted with experiments.

2.5 Solitons in lattices: discreteness effects.

In order to solve the discrete set of coupled nonlinear equations decribing the dynamics of the pendulum chain, we assumed that the interaction was strong enough to assure a smooth variation of θ_n from site to site. This allowed us to introduce a continuous function $\theta(x,t)$. This strong coupling condition is *essential to guarantee the existence of solitons moving freely in the system.* This can be understood from Fig. 1. If θ_n varies smoothly from 0 to 2π, all the values of θ are taken, including the value $\theta = \pi$ which corresponds to a pendulum pointing up, i.e. sitting on top of the potential due to gravitation. If one attempts to translate the soliton by one lattice spacing, the pendula in the core will rotate, some of them going down, i.e. decreasing their potential energy, and others going up, i.e. increasing their potential energy, until another pendulum points upward. The kink translation can be accomplished without spending energy because there is an exact balance between the energy gains and the energy losses of the different pendula. This is no longer true if the coupling string is weak with respect to the on-site potential due to gravitation. In this weak-coupling case the soliton becomes narrow with respect to the lattice spacing and its core contains only a few pendula as shown on Fig. 7 (Equation (7) shows that the soliton width is proportional to c_0^2, i.e. to the coupling constant C). The string is no longer strong enough to maintain a pendulum pointing upward. A soliton can only be in equilibrium if its center is situated *between* two lattice sites as shown in Fig. 7(a). The maximum of the gravitational potential, $\theta = \pi$, is no longer occupied by a pendulum. Figures 7(b) and (c) show what happens when one moves the soliton by one lattice spacing. During the process, one pendulum has to overcome the gravitational energy barrier by passing through the value $\theta = \pi$. Contrary to the strong coupling case, while this one climbs the potential barrier, no other pendulum goes down by the same amount. Therefore the intermediate position where one pendulum points upward (b) has more energy than the equilibrium positions (a) or (c). *In the discrete lattice there is a potential barrier for the translation of the soliton.* This "pinning potential" due to discreteness means that the soliton

146

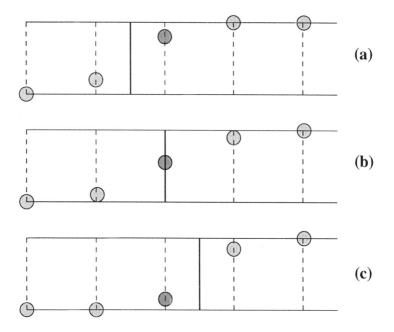

Figure 7: Schematic picture showing the propagation of a narrow soliton in the pendulum chain. The circles show the value of θ_n. For each figure the lower line corresponds to $\theta = 0$ and the higher line to $\theta = 2\pi$. The center of the soliton is indicated by the heavy vertical line. Fig. (a) shows the initial, stable, position of the kink in the chain. Fig. (b) corresponds to the intermediate state where the kink has moved by half a lattice spacing. A pendulum is in the upward position $\theta = \pi$. Fig. (c) is the new stable position after the kink has moved one site to the right.

can no longer be considered as a quasi-perticle moving freely on a flat potential energy surface. It should now be viewed as a particle moving on a washboard. The pinningg potential is well known in the theory of dislocations where it is called the Peierls-Nabarro potential. In more mathematical terms, one can say that the continuum equation is invariant by *any* translation $x \to x + \delta x$, while the discrete set of equations is only invariant by *discrete* translations $n \to n + 1$. Lattice discreteness has another consequence for a moving soliton. In the lattice the soliton no longer propagates at constant speed with a permanent profile. As shown on Fig. 8, the soliton is followed by a wake of small amplitude oscillations because the pendula which are not maintained by strong strings fall abruptly in the gravitational potential minimum and then oscillate around the minimum. The waves that the soliton leaves behind carry energy which is taken away from from its kinetic energy. Consequently, in the

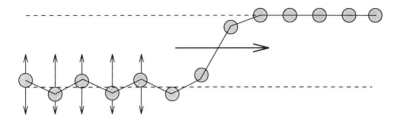

Figure 8: Shape of a narrow kink moving in a lattice. The pendula situated in the wake of the kink oscillate around their equilibrium position.

lattice, the soliton loses energy as if it were moving in a viscous medium. The consequence is that, even if it is launched with a high initial velocity, a narrow soliton eventually comes to rest unless it is driven by an external field. [11]

Therefore, in a discrete lattice, we must forget the picture of a freely moving soliton able to carry energy and information along, as a tireless messenger. This is particularly important for biological molecules. We have already said that they are highly deformable. It means that one segment of the molecule can easily move with respect to its neighbors. In other words it is generally weakly coupled to its neighbors. Thus solitons in biological molecules are likely to be narrow and pinned by lattice discreteness. Even though they do not move freely, they may generally diffuse under the influence of the thermal fluctuations and thus play an important role in the dynamics of the system.

Collective proton transport in hydrogen bonded chains.

Let us now consider a first example where the soliton concept provides some insight in a biophysical process, the protonic conductivity of hydrated proteins. Proton transport plays a fundamental role in cellular bioenergetics because pH gradients across membranes are the driving mechanism of many biomolecular reactions. In particular, proton transport is coupled to synthesis and hydrolysis of ATP which is the essential energy storage mechanism in biology. In spite of its importance, the mechanism of proton transport is far from being properly understood and this problem extends beyond biology. Even in physical systems which appear much simpler than biological molecules, such as ice, the mechanism of proton transport is not elucidated completely. The main question is the high proton mobility which is observed experimentally. When the proton-hydroxide permeability of biological membranes has been measured, it has been found to be orders of magnitude greater than expected from sodium-potassium permeabilities, [12] showing that protons play a special role. A similar observation had been made previously for ionic mobility in water solutions. [13]

This remark prompted Nagle and Morowitz[14,15] to suggest that the hydrogen bonded chains that were previously invoked by Bernal and Fowler[13] to explain proton transport in water and ice could also be responsible for proton transport in biology. Extending this idea, Antonchenko, Davydov and Zolotaryk[16] and Yomosa[17] proposed independently two models introducing the concept of *soliton* to describe collective effects in proton transport. These pioneering works have been followed by various models based on the same idea, but the experimental detection of the cooperative excitation had not yet been made. Recent experiments by G. Careri[18] who measured the conductivity of hydrated proteins revive the interest of these studies by providing a possible way to detect the cooperative effects. In this section, we review the soliton models and then propose an extension which attempts to combine the recent knowledge of interatomic potentials provided by ab-initio calculations, ideas from the soliton models, and a treatment of the thermal fluctuations to provide data that can be compared to these experimental results. What is the role of a *model* especially when it relies heavily on numerical simulations, as it is the case here? Our view is that it can help us to select among several explanations of the experiments, determine the basic mechanisms which dominate the experimental results, and make predictions to be tested in further experiments.

3.1   The physical problem and the first answers.

The first models for cooperative proton transport were intended to describe the very high mobility of protons in ice and water. In ice, the number of carriers is very low (about 10^{-6} per molecule at $-10°C$), but their mobility is very high. Measurements of the conductivity and saturation current indicate proton transfers from individual ions as frequent as 10^{13} per sec. or even more, comparable to $k_B T/h$ or even greater.[19] Moreover, as mentioned above, the theories of ionic mobility in water based on a picture of water as a fluid of definite viscosity and dielectric constant containing ions treated as spherical charged particles subject to resisting forces proportional to their velocity, which account satisfactorily for mobilities of large ions like K^+ or Cl^-, fail to account for the much larger mobilities of the H^+ or OH^- ions.[13]

These abnormal mobilities of the H^+ and OH^- ions, which are confined to water and related solvents, e.g. methyl alcohol, led Bernal and Fowler to propose the transport of protons along "filaments" as schematized on Fig. 9. These filaments are one-dimensional chains of water molecules connected by hydrogen bonds. Such hydrogen-bonded chains exist also in a variety of solid state systems such as ice, carbohydrates, lithium hydrazinium sulfate, imidazole[15] and form within the proteins that are embedded in biological membranes

As expected, Fig. 14(a) shows a general tendency to a higher mobility when temperature increases, but the finer shape of the curve is more interesting. First, in the case of protons, we found a dip in the mobility curve around $T = 290$ K which is yet unexplained. This structure does not seem to be a numerical artefact because the corresponding values have been checked by many extra calculations to improve the statistics and rule out the possibility of a simple fluctuation. This feature could perhaps be related to a similar effect observed by Nylund and Tsironis,[27] but it requires further studies. The mobility curve shows also clearly two regions. Below 150 K, mobility increases very fast with temperature while, above this temperature, a tendency to saturation is clearly observed. Although further analysis is necessary, we think that this behavior is an indication that a cooperative effect takes place and that the carrier mobility is assisted by the heavy-ion lattice deformation, in agreement with the picture provided by the static solutions shown in Fig. 13. This dynamical cooperative effect would lower significantly the barrier that the proton have to overcome as shown by the comparison of the Peirls-Nabarro barrier with mobile or fixed ions listed in table 1. To be efficient such a process requires a coherent motion of the heavy ions while the proton moves. This is only possible at low enough temperature. This would explain the saturation effect observed above 150 – 200 K.

This hypothesis is supported by the isotopic effect shown in Fig. 14(b). In the low temperature range the mobility of proton and deuterium are found to be similar. In fact we even find that the *heavier carrier is more mobile than the light one*, contrary to standard expectation. This would be understandable if, instead of the motion of an isolated proton, we were observing the motion of a "collective object" involving the proton and the surrounding heavy-ion distorsion. The effective mass of this object would have little to do with the mass of the individual proton or deuterium, hence the anomalous isotopic effect. Then, as temperature is raised above 150 K, the isotopic effect tends to recover the expected value of $\sqrt{2}$ for proton versus deuterium. This is consistent with the hypothesis discussed above that the collective effect is killed by the thermal fluctuations that break the coherence. The tendency toward the value $\sqrt{2}$ is however interrupted by the dip in proton mobility around 290 K that we discussed above.

Although this picture for the charge transport in the hydrogen-bonded chain is consistent with our numerical simulations, it certainly needs to be refined. And in particular the excess mobility of the deuterium has to be explained. It is possible that these heavier ions are more efficient than protons to drive the heavy-ion sublattice.

Although we have showed that the *soliton* picture is probably not exactly

appropriate to describe proton transport in hydrogen bonded chains becaus
the charge defect is so narrow that the soliton would be pinned by lattice ef
fects, some of the ideas contained in the soliton model seem to be be valid. I
particular, our numerical results with a model for which we have attempted t
reduce the arbitraryness of the parameter choice by using results of ab-initi
calculations, suggest that a *collective* effect involving a coherent distorsion o
the heavy-ion lattice could explain the fast increase of the proton mobilit
which is found. At higher temperatures, the collective effect seems howeve
to be destroyed by the thermal fluctuations. It is interesting that the result
obtained with this model are strongly supported by the results obtained inde
pendently by G. Careri and al. who have also observed an anomalous isotopi
effect in the low temperature range, which dispapears when temperature is in
creased.[18] The similarity is not perfect and, in particular, the anomaly in th
isotopic effect observed by G. Careri is much larger than the one we found. I
is too early to claim that this similarity provides a proof that collective effec
play a real role in proton transport, but we have perhaps now an indicatio
that this is the case. However there are other possibilities to explain an anoma
lous isotopic effect. Quantum effects have for instance been invoked to explai
thermally activated diffusion of hydrogen on tungsten surface[30] for which th
prefactor to the diffusion rate is higher for deuterium than for hydrogen. A
quantum effect is perhaps possible in the experimental system studied by G
Careri and coworkers, but it can be ruled out in our calculations which ar
purely classical.

4 The formation of solitons: nonlinear energy localization.

Solitons would be merely mathematical curiosities if their creation in a system
would require launching a wave with exactly the right profile. Experiments
with the pendulum chain show that solitons are easy to generate in a system
which meets the conditions for their existence. This is a general property
of soliton bearing systems. As shown in Fig. 15a, it is also very easy to
generate solitons in a wavetank by dropping a large stone at one end. The
initial disturbance evolves into one or several solitons due to the stable balance
between dispersion and nonlinearity as discussed above. This experiment is a
reduced model of the creation of a "tsunami" by an earthquake in the sea
The soliton can be considered as an attractor for localized perturbations in
the system. Sometimes the situation is more complicated because there are
several basins of attractions corresponding to different types of solitons that
can coexist in the system.

This is for instance the case for nonlinear electrical lines built as a sequence of filters including an inductance and a nonlinear capacitance made of a var-cap diode. [31] As shown in Fig. 15b, in such a device a triangular or square voltage pulse evolves into a pulse soliton well described by the KdV equation while a short wavetrain evolves into an envelope soliton well described by the Nonlinear Schrödinger equation. The oscillogrames shown in fig 15c show that the two types of solitons coexist in the system and can even pass through each other without being destroyed. This case illustrates an important property of nonlinear systems. The description of such a system by a nonlinear equation may not be unique and can depend on the particular type of excitation of interest. Even when only one equation can provide a good description of the system, the solutions of this equation can belong to different classes, showing the richness of nonlinear equations. Both the KdV equation and the NLS equation can be derived from the full equations of the electrical circuits, using different approximation schemes. And both provide a view of the system which is correct as long as the approximations that have been made to derive them stay valid.

In the cases that we have described up to now, solitons emerge from localized initial conditions. In these cases, the formation of the soliton is essentially a reshaping of the initial condition. But nonlinear phenomena can have a more important effect in physical and biological systems because they induce the *localization of energy* in the system. Nonlinear solitonlike excitations can also emerge from non-localized initial conditions or from thermal excitations. This property appeared from the first time in a numerical simulation performed in 1955 by Fermi, Pasta and Ulam (FPU) [32] but it was not recognized as such until 1965 because scientists we used to think in terms of linear excitations. In their work, FPU decided to investigate the behavior of a one-dimensional chain of 64 particles of mass m, interacting through forces that contain nonlinear terms, for times long compared to the characteristic periods of the corresponding linear problem. Their aim was to study "experimentally" the rate of approach to the equipartition of energy among the various degrees of freedom of the system. The hamiltonian of the chain is

$$H = \sum_{i=0}^{N-1} \frac{p_i^2}{2m} + \sum_{i=0}^{N-1} \frac{K}{2}(u_{i+1} - u_i)^2 + \frac{K\alpha}{3} \sum_{i=0}^{N-1} (u_{i+1} - u_i)^3 , \qquad (27)$$

where u_i and $p_i = m\dot{u}_i$ are the coordinate and momentum of the $i-th$ particle. They chose fixed boundary conditions $u_0 = u_N = 0$. The harmonic coupling constant is K, and α is a small parameter measuring the magnitude of the nonlinear term in the interaction potential. This hamiltonian can be expressed in terms of the normal coordinates A_k of the linearized hamiltonian

$$A_k = \sqrt{\frac{2}{N}} \sum_{i=0}^{N-1} u_i \sin\left(\frac{ik\pi}{N}\right) \tag{28}$$

as

$$H = \frac{1}{2}\left(\sum \dot{A}_k^2 + m\omega_k^2 A_k^2\right) + \alpha \sum C_{klp} A_k A_l A_p, \tag{29}$$

where $\omega_k = 2\sqrt{K/m}\sin(k\pi/2N)$ is the frequency of the $k - th$ normal mode and C_{klp} are constants. In the harmonic case ($\alpha = 0$), the energy stored initially in a given mode stays in that mode and the system does not approach thermal equilibrium. Fermi, Pasta, Ulam started their simulation by exciting the lowest mode ($k = 1$), i.e. by chosing a non localized initial condition having the shape of a plane wave with a wavelength equal to the size of the system. They thought that, for $\alpha \neq 0$, the mode coupling term in the expression (29) of the hamiltonian would cause energy to flow to the other modes, leading, in the long term to an equipartition of energy among the modes. At the beginning of the simulation, this is indeed what they observed: modes 2,3,4 became gradually excited.

But, at their surprise, after about 157 periods of the fundamental mode almost all the energy was back to the lowest mode. After this recurrence time the initial state was almost restored. Much longer calculations performed later showed that this periodic recurrence of the initial state could be repeated many times and a "super-recurrence" which restores the initial state almost perfectly was even found. This remarkable result, known as the FPU paradox, shows that introducing nonlinearity in a system does not guarantee an equipartition of energy. In order to understand the properties of the FPU system (and of most of the nonlinear systems), *it is essential to abandon the expansion on the linear modes* and to consider the full nonlinear excitations of the system. This was recognized by Zabusky and Kruskal in 1965 [33] who gave, ten years after the FPU discovery, an explanation of the FPU paradox. When one takes into account intrinsically the nonlinearity of the lattice, on gets an equation of motion which is *the KDV equation, i.e. the same equation that describes the propagation of hydrodynamic solitons in shallow water.* The derivation is a bit technical. It is given in Appendix A because it introduces a powerful method to analyze nonlinear systems, the *multiple scale expansion.*

It is important to notice that Zabusky and Kruskal solved the FPU paradox because they plotted the displacements in *real space* instead of looking at the Fourier modes. The formation of the solitons is associated to a *localization in space* of the energy of the initial signal which evolves into rather narrow solitons. The FPU recurrence is simply a manifestation of the stability of the

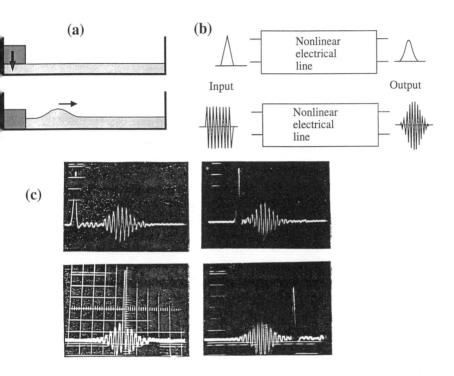

Figure 15: a) Generation of solitons in a wavetank from an arbitrary initial condition. (b) Two classes of initial conditions generating two different solitons in an electrical line: a triangular signal generates a pulse soliton while a short wave-train generates an envelope soliton. (c) Experimental collision of the two types of solitons in the electrical line.

solitons. The initial condition can be expanded in terms of its soliton content. The solitons that were not apparent in the initial profile show up clearly at a later time. Because they survive in the system, they can later come back almost to the same positions that they occupied initially, to form again a quasi-sinusoidal disturbance of the system. This is the recurrence of FPU.

The FPU simulation illustrates the localization of energy into pulse solitons. Another important phenomenon is the *modulational instability* of a plane wave in a nonlinear system. It is illustrated in Fig. 16 which shows the time evolution of a plane wave in a model electrical line.

Due to the nonlinear terms in the equation of propagation, a plane wave is unstable with respect to a modulation. A small amplitude modulation grows deeper and deeper and the wave spontaneously break up into wavepackets. The energy initially evenly distributed in the system concentrates itself into the wavepackets where the energy density can become significantly higher than in the initial wave. In the case presented in Fig. 16, the perturbation is simply caused by the wavefront and the modulation starts from the front before spreading along the wave. Appendix B shows how one can analyze the stability of a plane wave in the Nonlinear Schrödinger equation, introduced in section I-C, which describes for instance the electric field in a nonlinear optical fiber. The condition on the parameters of the equation so that it can have soliton solution is also the condition which leads to the modulational instability of a plane wave. Therefore we notice once again that, when an equation can sustain solitons, they are easily formed spontaneously in the system.

For application to biology, it is important to notice that nonlinear energy localization can also *occur from the thermal fluctuations* of the system. A thermalized nonlinear lattice can form spontaneously large amplitude localized oscillatory modes (i.e. breathers) which are extremely long lived. They can survive for thousands periods of the lattice oscillations. On timescales of millions of periods, there is an equipartition of energy in the lattice because breathers in contact with a thermal bath die out in one place and others form elsewhere, but during time intervals which are already long in comparison to typical periods of the dynamics of the lattice, the energy can self-localize in specific regions of the system. This could allow the large amplitude conformational changes which are necessary for some biological reactions without requiring a high temperature, because the thermal energy can be used very efficiently if it is essentially localized in the relevant place in the system. For instance the breathing of DNA, which can have important consequences when it traps small molecules that perturb the genetic code, is a an excitation which breaks temporarily several hydrogen bonds. This requires an energy which is well above the typical thermal energy $k_B T$. Nonlinear energy localization can

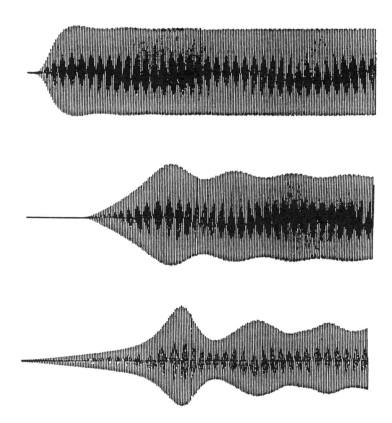

Figure 16: Evolution of a sinusoidal signal sent at the input of a nonlinear electrical line. The upper picture shows the signal at the beginning of the line, and the lower picture shows the signal after propagation in the nonlinear medium. The perturbation due to the front seeds up a self modulation of the signal.

perhaps provide a mechanism to explain this breathing.

5 DNA thermal denaturation: Nonlinear energy localization in a biological molecule.

DNA is central to all living beings, as it carries, for any species, all the information needed for birth, development, living and probably sets the average life-time of each species. Physically, DNA is a giant, double stranded linear macromolecule wound into the well known Watson-Crick double helix,[34] with length ranging from millimeters (bacteria) to meters (humans) and even approaching kilometers (salamander). Yet DNA is always packed so as to fit a micrometer size space, either as a loose structure in procaryotes (bacteria), or in densely packed, dedicated unit, the nucleus, in eucaryotes (the more evoluted cells).

The genetic information (or coding sequence) of a gene is always deposited on one strand only of the double helix. Gene expression involves successively transcription - the gene is transcribed from the DNA by an enzyme, RNApolymerase (RNAP), onto a "messenger" RNA - and translation, the process by which the messenger is translated into the gene product (protein) by a special "factory", the ribosome. Gene expression is carefully controlled and regulated, to ensure supply of the gene product in amounts at the time and the place required by the actual circumstances (intra- or extracellular).

DNA transcription is a typical example in which the dynamics of the molecule is essential to a biological function since the double helix has to be locally opened in order to expose the coding bases to chemical reactions. This process is however very complex because it is activated by an enzyme and it is probably still beyond a detailed analysis. Thermal denaturation has some similarities with the transcription because it starts locally by the formation of a so-called "denaturation bubble" similar to the local opening occurring in the transcription. Therefore investigating thermal denaturation is a valid preliminary step toward the understanding of the transcription. At temperature well below the denaturation temperature, DNA shows also large amplitude motions known as "fluctuational openings" in which base pairs open for a very short time and then close again. There is a rather clear experimental evidence that these very large amplitude motions are highly localized [35] and involve only few base-pairs at a time. They correspond to the local modes created by energy localization discussed in the previous section. These motions are important because, when the base pairs reclose, they can trap some external molecules causing a defect in the sequence. This process has been proposed as a possible mechanism of chemical carcinogenesis. [36] The fluctuational openings can be considered as intrinsic precursors to the denaturation.

The denaturation or "melting" transition is the separation of the two complementary strands. It can be induced by heating or by changing the ionicity of the solvent. It has been extensively investigated experimentally and models have been proposed to explain the complicated denaturation curves found in the experiments. [37] However these models are essentially Ising-like where a base pair is considered as a two-state system which is either closed or open. Such an approach cannot reproduce the full dynamics of the denaturation and it relies on phenomenological parameters for the probability of opening or cooperative character of this opening, which are not easily derivable from first principle calculations.

On another hand, the nonlinear lattice dynamics of DNA has recently been the subject of many investigations based on the idea that vibrational energy might be trapped into solitary wave excitations. This idea, originally suggested by Englander et al.[38] to explain the open states of the DNA molecule, has given rise to many investigations using soliton-like solutions to describe open states, transition between the A and B forms, or energy transport along the molecule. Most of these investigations have focused on the *propagation* of solitons along the double helix. However the biological function of DNA does not necessarily involve transport along the molecule. Consequently, although it is clear that a realistic model will exhibit nonlinear effects owing to the very large amplitude motions known to exist, their ability to propagate along the helix is not a requirement of the model. Rather, for denaturation (and transcription) we are concerned by the *formation and growth* of these excitations. It is this question that I want to discuss in these lectures.

5.1 A simple model for DNA melting and its statistical mechanics.

Let us consider an approach that goes further than the ising models, but still keeps the model as simple as possible in an attempt to determine the fundamental mechanism of the melting. Therefore we consider a simplified geometry for the DNA chain in which we have neglected the asymmetry of the molecule and we represent each strand by a set of point masses that correspond to the nucleotides. The characteristics of the model are the following:

(i) The longitudinal displacements are not considered because their typical amplitudes are significantly smaller than the amplitudes of the smaller transverse ones.[39] The stretching of a base-pair in the transverse direction is represented by a real variable y_n which can therefore describe all the states of the pair from closed ($y_n = 0$) to completely broken.

(ii) Two neighboring nucleotides of the same strands are connected by an harmonic potential to keep the model as simple as possible. On another hand, the bonds connecting the two bases belonging to different strands are extremely stretched when the double helix opens locally so that their nonlinearity must not be ignored. We use a Morse potential to represent the transverse interaction of the bases in a pair. It describes not only the hydrogen bonds but the repulsive interactions of the phosphate groups, partly screened by the surrounding solvent action as well. The hamiltonian of the model is then the following:

$$H = \sum_n \left[\frac{1}{2} m \dot{y}_n^2 + \frac{K}{2}(y_n - y_{n-1})^2 + D(e^{-a y_n} - 1)^2 \right] . \tag{30}$$

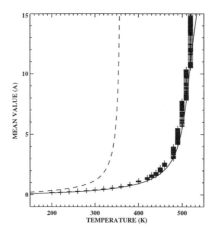

Figure 17: Variation of $\langle y \rangle$ versus temperature: the dash line corresponds to the TI results in the continuum limit, the solid line gives the exact TI results obtained by numerical solution of the TI operator, and the plus signs correspond to molecular dynamics simulations.

Since we are interested in the thermal denaturation transition of the molecule, the natural approach is to investigate the statistical mechanics of the model. For a one-dimensional chain containing N units with nearest neighbor coupling, the classical partition function can be calculated by the transfer integral operator (TI) method. [40,42]

The calculation is similar to the one performed by Krumhansl and Schrieffer[9] for the statistical mechanics of the ϕ^4 field. For the potential part, it yields $\mathcal{Z}_y = \exp(-N\beta\epsilon_0)$, where ϵ_0 is the lowest eigenvalue of the transfer operator. We can then compute the free energy of the model as $\mathcal{F} = -k_B T \ln \mathcal{Z} = -(Nk_B T/2)\ln(2\pi m k_B T) + N\epsilon_0$ and the specific heat $C_v = -T(\partial^2\mathcal{F}/\partial T^2)$. The quantity which gives a measure of the extent of the denaturation of the molecule is the mean stretching $\langle y_m \rangle$ of the hydrogen bonds, which can also be calculated with the transfer integral method. [41]

In the continuum limit approximation, the TI eigenvalue problem can be solved exactly, but experiments on proton exchange in DNA[35] show some evidence of exchange limited to a single base pair which suggests that discreteness effects can be extremely large in DNA. Therefore we have solved numerically the eigenvalue equation of the transfer operator without approximations. The TI operator is symmetrized and the integral is replaced by sums of discrete increments, using summation formulas at different orders. The problem is then equivalent to finding the eigenvalues and eigenvectors of a symmetric matrix.

Figure 17 compares the thermal evolution of $\langle y_m \rangle$ obtained with the continuum approximation and the exact numerical calculation, for the model parameters discussed in the next section. Both methods show a divergence of the hydrogen bond stretching over a given temperature, but the melting temperature given by the numerical treatment is significantly higher, pointing out the large role of discreteness in DNA dynamics if one uses realistic parameters for the model. The TI calculation shows that the specific heat has a broad maximum around the denaturation temperature.

5.2 Energy localization in the DNA molecule.

The thermodynamics of our DNA model shows that it exhibits a thermal evolution that is qualitatively similar to the denaturation of the molecule observed experimentally. But this statistical approach does not give informations on the *mechanism* of the denaturation, and in particular, does it start locally by the formation of denaturation bubbles in agreement with the experiments. In order to study this aspect, we have investigated the dynamics of the model in contact with a thermal bath by molecular dynamics simulation with the Nose method. [43] Most of the simulations have been performed with a chain of 256 base pairs with periodic boundary conditions, but in order to achieve better statistics, some simulations have been performed on a Connection Machine-200 with 16384 base pairs.

We have chosen a system of units adapted to the energy and time scales of the problem. Energies are expressed in eV, masses in atomic mass unit $(a.m.u.)$ and length in Å. The resulting time unit is $1 \ t.u. = 1.0214 \ 10^{-14}$s. The choice of appropriate model parameters is a very controversial topic, as attested by the debate in the literature. [44] There are well established force fields for molecular dynamics of biological molecules, but they have been designed to provide a good description of the *small* amplitude motions of the molecule and are not reliable for the very *large* amplitude motions involved in the denaturation. In our model, the Morse potential is an effective potential which links the two strands. It results from a combination of an attractive part due to the hydrogen bonds between two bases in a pair and the repulsive interaction between the charged phosphate groups on the two strands. The potential for the hydrogen bonds can be rather well estimated but the repulsive part is harder to determine because the repulsion is partly screened by ions of the solvent. Consequently we had to rely on estimations. The parameters that we use have been chosen to give realistic properties for the model in terms of vibrational frequencies, size of the open regions, etc, but future work will be needed to confirm our choice. We do not expect however that a better choice

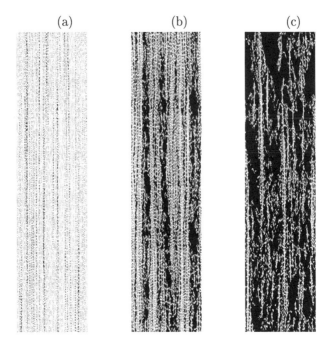

(a)　　　　　　　(b)　　　　　　　(c)

Figure 18: Results of molecular dynamics simulations at three different temperatures (a) $T = 150$ K, (b) $T = 340$ K, (c) $T = 450$ K. The horizontal axis indicates the position along the 256 cells of the molecule and the vertical axis indicates time. The stretching y_n of the base pairs along the molecule is indicated by a grey scale, the lighter grey corresponding to $y \leq -0.1$ Å and black indicating $y \geq 1$ Å. Therefore black regions show broken base-pairs.

would change *qualitatively* the results presented here. The parameters that we have chosen are : a dissociation energy $D = 0.04$ eV, a spatial scale factor of the Morse potential $a = 4.45$ Å^{-1}, a coupling constant $K = 0.06$ eV/Å, a mass $m = 300$ a.m.u. The constant of the Nose thermostat has been set to $M = 1000$.

Figure 18 shows a time evolution of the dynamics of the model at three temperatures. The stretching of the base pairs is indicated by a grey scale, darker dots corresponding to larger stretching.

Looking at this figure, one notices immediately two major features. First, as one moves along an horizontal direction, i.e. along the molecule for a given time, the amplitude of the stretching varies very much from site to site. This is especially true at high temperature, but it is still noticeable at 150 K, well below the melting temperature. This shows that there is *no equipartition o energy in this nonlinear system on a time-scale which is very long with respect to typical periods of the molecular motions*, but on the contrary a tendency for the energy to localize at some points which is more and more pronounced as temperature increases. However, the "hot-spots" due to nonlinear energy localization are dynamical entities. They, move, appear and die, and on a *macroscopic time-scale* one can consider that the average energy of all the sites is the same. At high temperature, the figure shows large black regions which correspond to denaturated regions of the molecule. These black areas are the denaturation bubbles observed experimentally. At the highest temperature shown here (Fig. 18 (c)) they extend over 20 to 50 base pairs and their boundaries are sharp. If the temperature is raised slightly above 540 K, the bubbles grow even more and finally extend over the whole chain: the molecule is completely denaturated.

The second remarkable feature on Fig. 18 can be observed by moving along a vertical line on the figure i.e. following the time evolution of a given base pair. If one choses one region of the molecule in which the energy is concentrated, one can see alternating black and light-grey dots. This is due to an *internal breathing* of the localized excitations that oscillate between a large amplitude (black dots in the figure) and a small amplitude state (light dots) in a regular manner. These motions are the fluctuational openings of DNA. They exist even well below the denaturation temperature and coexist with denaturated bubbles in the high temperature range. Figure 18 (b) shows that they play the role of precursor motions for the formation of the bubbles.

The calculation of the dynamical structure factor from the molecular dynamics results exhibits two types of excitations. In the high frequency range, one recognizes the phonon modes corresponding to linear motions of the chain. At low temperature their dispersion curve is well described by the linear dispersion curve resulting from the equations of motions of the model. Close to melting, on the contrary, most of the chain is on the plateau of the Morse potential and therefore experiences almost no restoring force that brings it back to $y = 0$. The dispersion curve is then the dispersion relation of a chain of harmonically coupled particles, without a substrate potential, i.e. a dispersion relation without gap. This is clearly observed in the molecular dynamics simulations and results in a phonon softening which should be observable experimentally in the vicinity of DNA melting transition. The second characteristic feature of the dynamical structure factor is a low frequency peak, associated to the fluctuational opening, which shifts to zero frequency as the denaturation bubbles form near the melting point.

The results of the molecular dynamics simulations illustrate the phenomenon of nonlinear energy localization discussed above and shows how it can drive a "melting transition". The localized "breathing modes" which are formed spontaneously in the system can be studied with standard methods of non-linear dynamics. They are approximately described by soliton solutions of a nonlinear Schrödinger equation. [45]

5.3 Is DNA melting a one-dimensional phase-transition?

The model discussed above has been able to describe some of the main features of DNA melting as it is observed experimentally. However there is a crucial point in which this model gives incorrect results, it is the *sharpness* of the phase transition. For an homopolymer, the experiments show that the melting occurs very abruptly over a temperature interval which is only a few K or even less. This poses a very fundamental question since DNA is basically a one dimensional system, which is not expected to have a phase transition. I want now to show that, *within a one-dimensional model with short range interactions*, a sharp transition is possible if one takes into account properly the nonlinearity of the base stacking interaction. [46] The possibility of a phase transition in one-dimensional DNA was already examined within the Ising-model approach by Poland and Scheraga [47] and Azbel [48] who concluded that it can be attributed to cooperativity effects and to the role of the winding entropy released when the two strands separate. Another view of this problem was proposed by Kosevich and Galkin [49] who showed that, for a double chain like DNA, one must not forget that the strands, when they are separated to form bubbles, explore a three-dimensional space. Therefore, although the model looks one-dimensional, when one counts the accessible states in the phase space this three-dimensional character appears. In our simple model, a corresponding feature comes from the Morse potential. When a portion of the chain is on the plateau of the Morse potential, it moves in fact in a two-dimensional space. Therefore the possibility of a phase transition cannot be ruled out. The interesting result is that it *does* exist provided the intra-strands interaction contain some nonlinearity. A simple extension of our DNA model (eq. 30) can describe the dynamics of these effects an gives a very sharp transition in agreement with the experiments. The stacking energy between two neighboring base pairs is described by the anharmonic potential:

$$W(y_n, y_{n-1}) = \frac{K}{2} \left(1 + \rho e^{-\alpha(y_n + y_{n-1})} \right) (y_n - y_{n-1})^2 . \qquad (31)$$

This new intersite coupling, replacing the simple harmonic coupling of our previous approach, is responsible for qualitatively different properties. The choice of this potential has been motivated by the observation that the stacking energy is not a property of *individual* bases, but a character of the base *pairs*

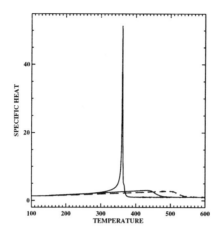

Figure 19: Variation of the specific heat versus temperature for the DNA model. The very narrow peak corresponds to the anharmonic coupling case ($\alpha = 0.35, \rho = 0.5$), the dotted curve and the solid broad peak to harmonic coupling ($k' = 1.5k$ and $k'' = k$, respectively).

themselves.[50] When the hydrogen bonds connecting the bases break, the electronic distribution on bases are modified, causing the stacking interaction with adjacent bases to decrease. In Eq. 31, this effect is enforced by the prefactor of the usual quadratic term $(y_n - y_{n-1})^2$. This prefactor depends on the *sum* of the stretchings of the two interacting base pairs and decreases from $\frac{1}{2}K(1 + \rho)$ to $\frac{1}{2}K$ when either one (or both) base pair is stretched. Although its form was chosen for analytical convenience, the qualitative features of potential 31 are in agreement with the properties of chemical bonds in DNA. They also provide the cooperativity effects that were introduced phenomenologically in the Ising models. A base pair that is in the vicinity of an open site has lower vibrational frequencies, which reduces its contribution to the free energy. Simultaneously a lower coupling along the strands gives the bases more freedom to move independently from each other, causing an entropy increase which drives a sharp transition.

Our approach can be compared to recent views on structural phase transition in elastic media which stress that intrinsic nonlinear features characterize the physics of these transformations, and extend the standard soft mode picture.[51] It is important to notice that, although cooperativity in introduced through purely *nearest neighbor* coupling terms, it has a remarkable effect on the 1D transition. Figure 19 shows the drastic change introduced by the anharmonic coupling on the specific heat of the model calculated by the transfer integral method. [46]

The full curve corresponding to the anharmonic stacking interaction shows a sharp peak very similar to that one would expect from a first-order phase transition, whereas the harmonic coupling investigated before gives only a smooth maximum. This result suggests that, although DNA structure is very complicated, a simple nonlinear model is able to reproduce with a good agreement the main experimental features of its dynamics for the fluctuational openings as well as the melting curves.

6 Conclusion.

It is interesting to notice that the original papers on solitons written by Scott Russel or Zabusky and Kruskal contained rather pictorial descriptions of the formation and dynamics of solitons. This is perhaps due to the unusual properties of the solitons for people who had been trained to think in terms of linear dispersive waves, so that they felt the need to explain in detail what they had seen to the sceptical reader. However, contrary to the title of our Section II-D it is certainly not correct to claim that "solitons are everywhere". In their enthusiasm for these excitations with exceptional properties, physicists have sometimes overestimated the role of solitons. It is nevertheless important to realize that nonlinear excitations provide *new tools* that it would be a pity to ignore in physics and biology. Using simple examples, we have shown how linear expansions can miss completely essential physical properties of a system. This is particularly characteristic for the pendulum chain. Soliton theory offers alternative methods. Multiple scale approximations, or expansion on a soliton basis, can be very useful to provide a description of some physical phenomena. Nonlinear energy localization is also a very important concept valid for a large variety of systems.

These concepts are probably even more relevant for biological molecules than for solid state physics because these molecules are very deformable objects where large amplitude motions or conformational changes are crucial for function. These motions are fundamentally nonlinear. To what extend the soliton concept is relevant in this context is not yet established, but it would certainly be a pity to leave it out of our toolbox. The remarkable stability of the solitons in the presence of perturbation is very similar to the exceptional ability of complicated biological processes to operate in an environment which is often modified by external factor and indeed subjected to thermal fluctuations. In both cases this stability is acheved through *collective effects*. It is perhaps this similarity that has pushed physicist to introduce the soliton concept in many areas of biology, *even when it is not valid!* (For instance some models rely on the mobility of the solitons although they are very narrow and would be trapped by discreteness). Thus it is probably good to end by a warning: models are useful to understand the basic processes in biophysics; solitons can help building some models *but* any model must be confronted to experiments and its validity cannot be established with *one* experimental results. This crucial point has been presented in a nice way by J. Krumhansl in the lecture he gave at the meeting "Nonlinear Excitations in Biomolecules" in Les Houches:

> I mentionned at the meeting, only partly in jest, what might be called "hot-air" models. Their basis is usually just circumstancial or semantic, and they would not survive experimental tests. As an example I took the question: "What makes the wind blow?" Answer: Note that whenever the wind blows the leaves on the tree move, complete correlation! The leaves derive energy from the sun by photosynthesis; they can oscillate and push the wind. Conclusion: photosynthesis makes the wind blow. Over and above the fact that cause and effect can never be obtained from correlation only. From the modeling point of view, any critical examination of this circumstancial logical chain would disclose multiple violations of the physics in the process proposed. Silly, yes. Nonetheless some purely circumstancial models, in physics and mathematics, of biomolecular behavior come close in their complete detachement from experimental reality.

Appendix A: Derivation of the KdV equation for a nonlinear lattice.

We consider a one-dimensional lattice of particles of mass m, coupled by the potential

$$V(r) = \frac{1}{2}K_2 r^2 + \frac{1}{3}K_3 r^3 + \frac{1}{4}K_4 r^4 , \qquad (32)$$

where r is the interatomic distance. Denoting by $u_i(t)$ the displacement of atom i with respect to its equilibrium position, its equation of motion is

$$\begin{aligned} m\ddot{u}_i \; = \; & K_2(u_{i+1} + u_{i-1} - 2u_i) \\ & + K_3 \left[(u_{i+1} - u_i)^2 - (u_i - u_{i-1})^2 \right] \\ & + K_4 \left[(u_{i+1} - u_i)^3 - (u_i - u_{i-1})^3 \right] \end{aligned} \qquad (33)$$

If we introduce the dimensionless displacement v by $u/h = av$ where h is the lattice spacing, and $a \ll 1$ gives the scale of the displacement, we get

$$\ddot{v}_i = \frac{K_2}{m} \left\{ \begin{array}{l} (v_{i+1} + v_{i-1} - 2v_i) \\[2ex] + \dfrac{aK_3h}{K_2} \left[(v_{i+1} - v_i)^2 - (v_i - v_{i-1})^2 \right] \\[2ex] + \dfrac{a^2 K_4 h^2}{K_2} \left[(v_{i+1} - v_i)^3 - (v_i - v_{i-1})^3 \right] \end{array} \right\} . \tag{34}$$

We are interested in displacements which show only a small variation from one site to the next, which allows us to use a continuum limit approximation, i.e. replace the discrete set of functions $v_i(t)$ by a function $v(x,t)$ depending continuously on space as well as time. Let us introduce dimensionless space and time variables by

$$X = \frac{\epsilon x}{h} \qquad \theta = \frac{ct}{h} \tag{35}$$

where c defined by $K_2/m = c^2/h^2$, which has the dimensions of a velocity, is the speed of sound (long wavelength linear excitations of the lattice) and $\epsilon \ll 1$ is introduced to express in a mathematical sense the slow spatial variation of the fuction v. In terms of these new variables, $\partial^2 v/\partial t^2$ becomes $(c^2/h^2)(\partial^2 v(X,\theta)/\partial\theta)$, and $v_{i\pm1}(\theta)$ is expressed with a Taylor expansion of $v(X,\theta)$ around $X_i = ih$

$$\begin{aligned} v_{i\pm1}(\theta) &= v[\epsilon(\frac{x_i \pm h}{h}), \theta] = v(X_i \pm \epsilon, \theta) \\ &= v(X_i, \theta) \pm \epsilon \frac{\partial v}{\partial X}(X_i) \frac{1}{2}\epsilon^2 \frac{\partial^2 v}{\partial X^2}(X_i) \\ &\pm \frac{1}{6}\epsilon^3 \frac{\partial^3 v}{\partial X^3}(X_i) + \frac{1}{24}\epsilon^4 \frac{\partial^4 v}{\partial X^4}(X_i) + O(\epsilon^5) . \end{aligned} \tag{36}$$

Consequently we have

$$\begin{aligned} v_{i+1}(\theta) + v_{i-1}(\theta) - 2v_i(\theta) = \\ \epsilon^2 \frac{\partial^2 v}{\partial X^2}(X_i) + \frac{\epsilon^4}{12}\frac{\partial^4 v}{\partial X^4}(X_i) + O(\epsilon^6) , \end{aligned} \tag{37}$$

and, from

$$(v_{i+1} - v_i) = \epsilon^2 \frac{\partial v}{\partial X}(X_i) + O(\epsilon^3) \tag{38}$$

$$(v_i - v_{i-1}) \;=\; \epsilon^2 \frac{\partial v}{\partial X}(X_{i-1}) + O(\epsilon^3) \tag{39}$$

with a second Taylor expansion for the derivatives, we get

$$(v_{i+1} - v_i)^2 - (v_i - v_{i-1})^2 =$$

$$\epsilon^2 \left[\epsilon \frac{\partial}{\partial X} \left(\frac{\partial v}{\partial X}(X_i) \right)^2 + O(\epsilon^2) \right]$$

$$= \epsilon^3 \frac{\partial}{\partial X} \left[\frac{\partial v}{\partial X}(X_i) \right]^2 + O(\epsilon^4) \tag{40}$$

Similarly we get

$$(v_{i+1} - v_i)^3 - (v_i - v_{i-1})^3 = \epsilon^4 \frac{\partial}{\partial X} \left[\frac{\partial v}{\partial X}(X_i) \right]^3 + O(\epsilon^5) . \tag{41}$$

As explained in the text, we assume that the third order nonlinearity is small and define $K_3 h / K_2 = \epsilon p / 2$, while the quartic nonlinearity is taken of order 1 by setting $K_4 h^2 / K_2 = q/3$. Introducing the different expressions calculated above into eq. (34), keeping only the terms up to order ϵ^4, and dividing by $\epsilon^2 / h^2 = K_2/m$, we obtain

$$\frac{\partial^2 v}{\partial \theta^2} \;=\; \epsilon^2 \frac{\partial^2 v}{\partial X^2} \left[1 + \epsilon^2 a p \frac{\partial v}{\partial X} + \epsilon^2 a^2 q \left(\frac{\partial v}{\partial X} \right)^2 \right]$$

$$+ \frac{\epsilon^4}{12} \frac{\partial^4 v}{\partial X^4} + O(\epsilon^5) \tag{42}$$

This equation is still difficult to solve, but it can be put into a simpler form by changing to the frame moving at the speed of sound c. In the absence of dispersion and nonlinearity, any deformation v propagates in the lattice at the speed of sound and therefore it appears as static in the frame which moves at velocity c. Here we consider a case with weak dispersion and weak nonlinearity. This can been seen from eq. (42) because the dispersive and nonlinear terms are of order ϵ^4. Therefore, we can expect that a deformation v will change very slowly in the frame moving at the speed of sound. This can be made quantitative by looking for a solution which is function of a "slow time" $\tau = \epsilon^3 \theta$. The change of frame means that the space variable is $\xi = (\epsilon/h)(x - ct) = X - \epsilon\theta$, and the space and time derivatives are modified accordingly

$$\frac{\partial}{\partial X} \;=\; \frac{\partial}{\partial \xi} \tag{43}$$

$$\frac{\partial^2}{\partial\theta^2} = \epsilon^6\frac{\partial^2}{\partial\tau^2} - 2\epsilon^4\frac{\partial^2}{\partial\xi\partial\tau} + \epsilon^2\frac{\partial^2}{\partial\xi^2} . \tag{44}$$

If these expressions are introduced into eq. (42) for which only terms up to order 4 must be conserved since we have already dropped some terms of the order ϵ^5, we find that the term $v_{\xi\xi}$, which is of order ϵ^2 cancels on both sides and, if we define $w = v_\xi$ the remaining terms, which are all of order ϵ^4 give the equation

$$w_\tau + \frac{1}{2}apww_\xi + \frac{1}{2}qa^2w^2w_\xi + \frac{1}{24}w_{\xi\xi\xi} = 0 , \tag{45}$$

which is a generalization of the KdV equation known as "modified KdV equation". If we consider an interatomic potential (32) without the quartic term as for the FPU problem, i.e. we set $q = 0$, eq. (45) reduces to the KdV equation. The final equation is not invariant with respect to time reversal while the initial one was, but one must keep in mind that eq. (45) is written in a frame moving at the speed of sound c. In this frame, a solution with positive velocity is a supersonic signal in the original lattice, while a solution with a negative velocity is a subsonic signal in the lattice. The KdV solitons are *always supersonic* signals.

Appendix B: Modulational Instability in the NLS equation.

Let us start from the NLS equation

$$i\frac{\partial\psi}{\partial t} + P\frac{\partial^2\psi}{\partial x^2} + Q|\psi|^2\psi = 0 \tag{46}$$

The plane wave

$$\psi = A_0 e^{i(\kappa x - \Omega t)} \tag{47}$$

is a solution of eq. (46) if $\Omega = P\kappa^2 - QA_0^2$. In order to study the stability of this solution we look for the time evolution of a small perturbation of its amplitude or phase. This is done by chosing

$$\psi(x,t) = [A_0 + b(x,t)]e^{i[\Omega t + \kappa x + \theta(x,t)]} = [A_0 + b(x,t)]e^{i\Phi} , \tag{48}$$

where $b(x,t)$ and $\theta(x,t)$ are supposed sufficiently small so that we can keep only the expressions which are firts order in b and *theta*. Therefore we have

$$\frac{\partial\psi}{\partial t} = \frac{\partial b}{\partial t}e^{i\Phi} + i\Omega(A_0 + b)e^{i\Phi} + i\frac{\partial\theta}{\partial t}e^{i\Phi} + O(b^2, \theta^2, b\theta)$$

$$\frac{\partial^2\psi}{\partial x^2} = \frac{\partial^2 b}{\partial x^2}e^{i\Phi} + i\kappa\frac{\partial b}{\partial x}e^{i\Phi} - \kappa^2(A_0 + b)e^{i\Phi} - \kappa\frac{\partial\theta}{\partial x}A_0 e^{i\Phi}$$

19. Mou-Shan Chen, L. Onsager, J. Bonner and J. Nagle *Hopping of ions in ice* J. Chem Phys. **60**, 405-419 (1974).

20. N. Bjerrum, *Structure and properties of ice,* Science **115**, 385-390 (1952).

21. J.F. Nagle, *Theory of Passive proton Conductance in Lipid Bylayers.* J. of Bioenergetics and Biomembranes **19**, 413-426 (1987).

22. Xiaofeng Duan and S.Scheiner, *Modeling of coupled proton transfers by analytic functions* International Journal of Quantum Chemistry: Quantum Biology Symposium **19**, 109-124 (1992).

23. N.D. Sokolov, M.V. Vener and V.A. Savel'ev, *Tentative study of strong hydrogen bond dynamics: Part II Vibrational frequencies considerations.* J. of Molecular Structure **222**, 365-386 (1990).

24. M. Peyrard, St. Pnevmatikos, and N. Flytzanis, *Dynamics of two-component solitary waves in hydrogen-bonded chains,* Phys. Rev. A **36**, 903-914 (1987).

25. J. Halding and P.S. Lomdahl, *Two-component soliton model for proton transport in hydrogen-bonded molecular chains,* Phys. Rev. A **37**, 2608-2613 (1998).

26. A.V. Savin and A.V. Zolotaryuk, *Dynamics of ionic defects and lattice solitons in a thermalized hydrogen-bonded chain,* Phys. Rev. A **44**, 8167-8183 (1991).

27. E. Nylund and G. Tsironis, *Evidence for solitons in hydrogen-bonded systems,* Phys. Rev. Lett. **66**, 1886-1889 (1991) and E. Nylund, K. Lindenberg and G. Tsironis *Proton dynamics in hydrogen-bonded systems* J. of Statistical Physics **70**, 163-181 (1993).

28. G. Tsironis, *Proton-solitons bridge physics with biology,* in Nonlinear Excitations in Biomolecules, M. Peyrard, editor (Les Editions de Physique-Springer Verlag, Paris 1995).

29. W.G. Hoover, *Canonical dynamics: Equilibrium phase-space distributions.* Phys. Rev. **A31**, 1695-1697 (1985).

30. K.F. Freed *Quantum mechanical theory of isotope effect on thermally activated hydrogen migration on W(110),* J. Chem Phys. **82**, 5264-5268 (1985).

31. M. Remoissenet, *Waves called solitons: concepts and experiments,* (Springer Verlag, Berlin 1994)

32. The Fermi Pasta Ulam work was never published as a paper because Fermi died before the paper was written. It appeared as a Los Alamos report which was later included in the collected works of Fermi. A reprinted version can be found in the book *The Many Body Problem,* D.C. Mattis, editor (World Scientific, Singapore 1993).

33. N.J. Zabusky and M.D. Kruskal, *Interaction of "solitons" in a collision-*

less plasma and the recurrence of initial states, Phys. Rev. Lett. **15** 240-243 (1965).

34. WŚaenger, in *Principles of Nucleic Acid Structure* (Springer Verlag Berlin, Heidelberg, New-York, Tokyo 1984).

35. M. Guéron, M. Kochoyan and J.L. Leroy, *A single mode of DNA base pair opening drives imino proton exchange,* Nature **328**, 89 (1987)

36. J. Ladik and J. Cizek, *probable physical mechanisms of the activation o oncogenes through carcinogens,* Int. J. of Quantum Chemistry XXVI, 95£ (1984).

37. R.M. Wartell and A.S. Benight, *thermal denaturation of DNA molecules a comparison of theory with experiments,* Physics Reports **126**, 67 (1985)

38. S.W. Englander, N.R. Kallenbach, A.J. Heeger, J.A. Krumhansl and S Litwin, *Nature of the open states in long polynucleotide double helices Possibility of soliton excitations,* Proc. Nat. Acad. Sci. USA, **777**, 722? (1980)

39. J.A. MacCammon, S.C. Harvey, in *Dynamics of proteins and nucleic acids* (Cambridge University Press, Cambridge 1988).

40. M. Peyrard and A.R. Bishop, *Statistical mechanics of a nonlinear mode for DNA denaturation,* Phys. Rev. Lett **62**, 2755 (1989).

41. D. J. Scalapino, MŚears and R. A. Ferrel, *Statistical mechanics of one-dimensional Ginzburg-Landau fields,* Phys. Rev. B **6**, 3409 (1972).

42. J. F. Currie, J. A. Krumhansl, A. R. Bishop, S. E. Trullinger, *Statistica mechanics of one-dimensional solitary-wave-bearing scalar fields: exac results and ideal-gas phenomenology* Phys. Rev. B **22**, 477 (1980).

43. S. Nose, *A unified formulation of the constant temperature molecular dynamics methods,* J. Chem. Phys. **81**, 511 (1984)

44. L.L. Van Zandt, *DNA solitons with realistic parameters,* Phys. Rev. A **40**, 6134 (1989), M. Techera, L.L. Daemen, E.W. Prohofsky, *Commen on "DNA solitons with realistic parameters",* Phys. Rev. A **42**, 5033 (1990), L.L. Van Zandt, *Reply to "Comment on DNA solitons with realistic parameters",* Phys. Rev. A **42**, 5036 (1990), and references therein.

45. T. Dauxois, M. Peyrard and C.R. Willis, *Localized breather-like solutions in a discrete Klein-Gordon model and application to DNA,* Physica D **57**, 267 (1992).

46. T. Dauxois and MPeyrard, *Entropy driven DNA denaturation,* Phys. Rev. E **47**, R44 (1993).

47. D. Poland, H. Scheraga, *Phase transitions in one-dimension and the helix-coil transition in poly-amino acids,* J. Chem. Phys **45**, 1456 (1966).

48. M.Ya. Azbel, *Generalized one-dimensional Ising model for polymer thermodynamics* J. Chem. Phys. **62**, 3635 (1975).

49. A.M. Kosevich and V.L. Galkin, *Phase transition in a double polymer chain in an external field,* Soviet Physics JETP **33**, 444 (1971).

50. C. Reiss, private communication (1991).

51. W. C. Kerr, A.M. Hawthorne, R.J. Gooding, A.R. Bishop and J.A. Krumhansl, *First-order displacive structural phase transition studied by computer simulation* Phys. Rev. B **45**, 7036 (1992).

52. J.P. Boucher, *Nonlinear excitations in antiferromagnetic chains,* Hyperfine Interactions **49**, 423 (1989) and J.P. Boucher, L.P. Regnault, J. Rossat-Mignod, Y. Henry, J. Bouillot and W.G. Stirling, *Solitons in the paramagnetic and partially disordered phases of $CsCoCl_3$,* Phys. Rev. B **31**, 3015 (1985).

53. A.M. Kosevich and A.S. Kovalev, *Self-localization of vibrations in a one-dimensional anharmonic chain,* Sov. Phys. JETP **40**, 891 (1975).

54. S. Takeno and A.J. Sievers, *Intrinsic localized modes in anharmonic crystals,* Phys. Rev. Lett., **61**, 970 (1988).

55. J.A. Krumhansl and J.R. Schrieffer, *Dynamics and statistical mechanics of a one-dimensional hamiltonian for structural phase transitions,,* Phys. Rev. B **11**, 3535 (1975).

THE IMMUNE SYSTEM, AND WHY MODELING IT MAKES SENSE

FRANCO CELADA

Cattedra di Immunologia, Università di Genova
and Hospital for Joint Diseases, NYU, New York

Past and present

The immunologist's view of the immune system has been gradually shifting during the last decades. Before, there were a number of responses, nonspecific and specific, humoral (i.e., mediated by antibodies) and cellular (where the immunocytes themselves are the effector agents). Their power was tremendously enhanced by memory (mechanisms by which the response to an already experienced insult is larger, quicker and better than the first time) and limited by tolerance (the provision by which a response is waived if it may hurt the organism itself). The responses were triggered and controlled by the stimulus. A pushed button, that – through a number of mechanistic steps – yielded a predictable result.

The new picture, bearing some likeness of a nervous system, did materialize through an impressive body of data on regulation by means of 2-cell, 3-cell, cell cluster interactions and a new synapse-like environment where receptors meet.

Two fundamental features have been identified, which are the basis of *humoral* and *cellular* responses respectively:

a) the cell-cell cooperation (between Th and macrophages and between Th and B cells made possible by the binding, internalizing, processing and presentation of the antigen by one partner (APC) and the secretion of an array of lymphokines (activating or regulating molecules with a short range/time activity) by the other (the Th cell);

b) the phenomenon of MHC restriction, by which the T cell mediated cytotoxicity occurs under condition that the targeted antigen is presented on self MHC class I receptors.

2 The unconscious brain

The immune system behaves more like a cognitive system (the unconscious mind that manages the organism knowledge of the exterior chemical and bio-

logical world) than a automatic or reflex response machinery. Thanks to cellu
lar consultations and cooperations, each response (or lack of it) is preceded b
a time devoted to the evaluation of the stimulus and the circumstances befor
deciding to respond or not.

There are clear analogies between the behavior of the human mind con
fronted - in a classic example discussed by Umberto Eco - with an intrinsicall
ambiguous message [1] and the initial steps of the humoral response. In th
predicament, the mind constructs alternative interpretations but only decide
which one to adopt when it obtains independent additional information. A
B lymphocyte, through its antibody receptors, makes contact with a speci
ically binding antigen; then, instead of immediately going through the step
that would start the synthesis and secretion of antibodies, it hesitates, collect
information about parts of the antigen distinct from the determinant it ha
bound, processes these parts, displays the resulting peptides on ad hoc mem
brane receptors, and waits for a T cell to recognize them and give a *go* signa
However, the *go* signal may never come: typically when the T cell in questio
does not exist any more, having been eliminated in the thymus as a dangerou
self aggressive cell. Therefore, the lack of signal signifies that the antigen i
not foreign but self: it would be a self aggression to launch an antibody attac
against it. Probably, both the mind's and the immune system's behaviors ar
the product of trials and selections that took place respectively during specie
development and philogenesis, and are now embodied in their respective actin
codes.

3 The Principles

The principles embodied in the immune system's code are directed at th
survival of the organism and of the immune system itself. The mode is Dar
winism at the level of the individual organism. In order to be able to fulfill thi
formidable assignment, the immune system has generated a complexity muc
larger than was imagined in the 70's. Still, it is features discovered then tha
are in the center of the organization emerging now.

The first principle is: **Be prepared for any possible foreign invader**
This is mainly achieved by building an almost infinite repertoire of recepto
specificities. The second principle is: **Do not inflict damage to self, di
rectly or indirectly**. These two principles are in a *Catch-22* relation since b
enhancing the repertoire of specificities and the power of the effectors there i
an automatic increase in probability of autoimmune aggression by sheer cross
reaction. The evolutionary solution of this paradox was reached by separatin
effector power from decisional power, assigning them to different cell lineage

respondence with the phenomenon. At that point, the simulation may project a reconstruction of nature which is accessible in all its facets, easy to observe and to measure, with temporal and causal relationships neatly defined and dynamic attractors earmarked. This is the basis of the use of models as a didactic tool (e.g., in teaching Immunology).[11] Arguably even more important is their function in the very motor of scientific progress, the generation of hypotheses/abductions. The springboard for imagination can be as well - or even better - clear and quantitative simulation, as the natural phenomenon is so often concealed or cryptic.

) **Models as testing instruments**

The proper test of a hypothesis is the experiment. It consists of reproducing a natural phenomenon under controlled conditions and comparing its dimensions with those predicted by the prevailing theory. To control the experimental conditions means to be able to vary all parameters, one at a time, in order to assess, by state of the art instruments, their effects on the results.

In Immunology as in all biosciences, experiments have become more and more sophisticated and thus remote from *in vivo* observations. Cell populations, single cells and subcellular systems are studied *in vitro*, or in intermediate systems such as adoptive transfers and transgenic animals. A further extension step along this tendency is the performance of experiments *in machina*, that is, using dynamic models of the system to be tested. This kind of experimentation has total flexibility and the loss of reality due to abstraction is not qualitatively different than the loss suffered between *in vivo* and *in vitro*. However, a third stride is taken, and there is a give-and-take between abstraction and demonstrativeness. For this reason, experiments *in machina* cannot substitute for *in vivo* or *in vitro* tests. But, if the parameters introduced are realistic, the results of the simulation can give indications on which directions would be more or less promising to probe with bench experiments. This is a real help in the dynamics of the grey layer of Planet Immunology, and here experiments in machina can save time, money and scores of laboratory animals.

As opposite hypotheses may and do coexist for long periods of time before a definitive experiment can be devised and performed, modeling has been applied as a preliminary clarifying and testing tool. For this purpose, the model must be neutral enough to be able to simulate both alternative postulates on the table. The primary aim of this exercise is to reveal the discriminating parameters on which biological experiments may be

194

fruitfully aimed. An example of this use is the modeling of B-cell hy permutation after primary antigenic stimulation. *In vivo* mutations ar found with significantly higher frequencies on the gene segments codin for CDRs (the Complementarity Determining Regions of the antibody' One theory assumes an even distribution of mutations over the entir V-region and attributes the skewed distribution observed to the negativ selection of mutants suffering residue changes in sensitive positions out side CDRs. [12] The alternative theory postulates that mutational event are originally confined, or at least strongly focused to the DNA sequence coding for CDRs, where residue changes have no disruptive effects on th antibody architecture. [13]

The modeling effort consisted in simulating both hypotheses - in a cellula automaton - and comparing the effect of increasing mutation frequencie (another biological unknown) in the two conditions, in terms of affinit maturation of the response. Clearly, it was not possible to decide whic hypothesis should be preferred, but as a result of our simulations, i was found the most discriminating parameter is the cloud growth rat of specific B cells, which - during the secondary response is up to thre hundred times faster for focused mutations than for randomly distribute events. This indication should help the final solution of the problem b ad hoc designed biological experiments.

ACKNOWLEDGEMENTS. The author's work discussed in this article wa supported by grants from The National Institutes of Health (1 RO1 AI33660 04A1 and 1 RO1 AI 42262-01) and from the Center for Alternatives to Anima Testing (# 97023 and 9827) of the Johns Hopkins University).

References

1. U. Eco in *The Semiotics of Cellular Communication in the Immune Sys tem*, F. Celada, N.A. Mitchison, E.E. Sercarz, T. Tada, Editors (Springe Verlag , Berlin and Heidelberg 1988).
2. P. Lake and N.A. Mitchison, *Regulatory mechanisms in the immune re sponse to cell surface antigens*, Cold Spring Harbor Symp. Quant. Biol **41** 589 (1976).
3. E. Roosnek and A. Lanzavecchia *Efficient and selective presentatio of antigen-antibody complexes by rheumatoid factor B cells*, J. of Exp Medicine. *173* 487-489 (1991).
4. F. Manca, E. Seravalli, M.T. Valle, D. Fenoglio, A. Kunkl, G. Li Pira S. Zolla-Pazner and F. Celada , *Non-covalent complexes of HIV gp12 with CD4 and/or mAbs enhance activation of gp120-specific T clone*

and provide intermolecular help for anti-CD4 antibody production, Intl. Immunology. **5** 1109-1117 (1993).

5. P. Matzinger, *Tolerance, danger, and the extended family*, Ann. Rev. Immunol. **12** 991-1045 (1994).

6. M. Cella, F. Sallusto and A. Lanzavecchia A, *Origin, maturation and antigen presenting function of dendritic cells*, Curr. Opin. Immunol. (in press 1997).

7. F. Celada, *Bricolage in cellular immunity*, Intl. Rev. Immunol. **4** 107-194 (1989).

8. F. Celada and P.E. Seiden, *Affinity maturation and hypermutation in a simulation of the humoral immune response*, Eur. J. Immunol. **26**, 1350-1358 (1996).

9. C.S. Peirce, Collected Papers, Vol 1:1931-1958, Hartshoene C, Weiss P, Editors (Harvard University Press, Cambridge, MA 1968).

10. K.R. Popper *Logik der Forschung* (Julius Springer Verlag, Vienna 1935).

11. F. Celada and P.E. Seiden, *Teaching Immunology: A Montessori approach using a computer model of the immune system*, In *T Lymphocytes: Structure, Functions, Choices*, F. Celada and B. Pernis, Editors, NATO ASI Series, Vol. A 233 (Plenum Press, New York and London 1992), pp. 215-225.

12. M.J. Shlomchik, S. Litwin and M. Weigert, in *Progress in Immunology*, F. Melchers, Editor (Springer Verlag, Berlin 1989), p. 415.

13. A.G. Betz, M.S. Neubauer and C. Milstein C, Immunol. Today **14** 405 (1993).

IMMUNE SYSTEM MODELLING

MICHELE BEZZI

*Dipartimento di Fisica, Università di Bologna, Via Irnerio 46,
40126 Bologna, Italy, and INFN , Bologna, Italy*

1 Introduction

In this paper we present a short survey of some results of theoretical immunology. This paper is not an exaustive report of models of immune systems, but only an arbitrary selection of some of them. The sketch of the paper is as follows: in the first section we will give a short description of some remarkable features of immune systems, [a] in section 2 we describe some models of immune response and of idiotypic network; in section 3 we describe a microscopic model for humoral response proposed by Celada and Seiden.

2 Immune System, cardinal features

The defense mechanisms used by the body against an attack from foreign substances (*antigens*) are several, they include: physical barriers, phagocytic cells, different clones of particular white blood cells (*lymphocytes*) and various blood-borne molecules (e.g., *antibodies*). Some of these mechanisms are present prior to exposure to antigens and their action is non-specific, i.e., they do not discriminate between different antigens and their response doesn't change upon further exposure to the same antigen. This kind of response is called *natural immunity*. There are other mechanisms with more specific behavior, they are induced by antigens and their response increases in magnitude and defense capabilities with successive exposures to the same antigens. These mechanisms are called *acquired* (or *specific*) *immunity* and are what we consider in this paper.

The main features of specific immunity are:

- *Specificity*: the immune response is highly specific for distinct antigens, that is only a small number of particular membrane receptors recognize a given antigen. These receptors bind specific portions of the antigen (called *antigenic determinants* or *epitopes*). Only a small fraction of the system recognizes any particular epitope, and cells which do recognize

---

[a] A complete description of immune system features can be found in any immunology textbook, e.g. Abba *et al.* (1991). [1]

antigenic epitopes are positively selected by this recognition process so that their population increases (*clonal selection*).

- *Diversity*: the possible number of lymphocytes with different specificity (different *clones*) is extremely large ($\geq 10^9$), although the total diversity actually existing in an animal is much smaller. The diversity is due to the variability of lymphocyte receptors, which derives from the mechanism of genetic recombination that occurs during cell development.

- *Maturation*: the response evolves by increasing the average affinity to the antigen through competition and selection for binding following mutation of the genes coding for the B-cell receptor.

- *Memory*: exposure of the immune system to an antigen enhances, in quality and quantity, its capability to respond to a second exposure to the same antigen (*secondary response*). This is due mainly to the persistence, after any response, of re-stimulable cells (memory cells) ready to mount a new response to the same antigen.

- *Discrimination of self from non-self*: the immune system doesn't respond to substances produced by the human body (*tolerance to self*).

The specificity of the immune response is due to a class of white blood cells called lymphocytes. These are not able to begin a response without the help of various types of cells known as *antigen presenting cells* (APC), such as macrophages, dendritic cells, etc. Lymphocytes are present in the blood, lymph and lymph nodes; in the human body there are about 10^{10} lymphocytes. The two major classes are: *B* and *T lymphocytes*.

In mammals B lymphocytes (so called because in birds they are produced in an organ called the *Bursa Fabricii*) mature in the bone marrow, they are able to bind antigen and to produce antibodies with the same specificity as their membrane receptors. Upon binding antigen a B cell will endocytose the antigen into small pieces between about 8 to 15 amino acids long (peptides). These pieces are bound to surface receptors called major histocompatibility molecules (MHC class II). Helper T cells are able to recognize and bind to these MHC/peptide complexes. When they do they trigger T-and B-cell proliferation and differentiation into memory cells and *plasma cells* (which produce antibodies).

T lymphocytes are born in the bone marrow, and then migrate to the thymus where they mature and are selected. Here a lot of T lymphocytes are destroyed to avoid auto-immune responses (*clonal deletion*). There are two main different types of T lymphocytes.

- *Helper T cells* (Th) which are involved in B-cell activation; they have receptors that can bind MHC/peptide complexes. As a consequence of this process, they produce a set of particular molecules, called *cytokines*(e.g., interleukin 2 and 4), that activate B and T cells;

- *Cytotoxic T lymphocytes* also called *killer T cells* (CTL or Tk) which can recognize, in an antigen specific way, cells infected by viruses, and kill them by lysis. The recognition is similar to that of helper T cells in that pieces of antigen are presented on another type of MHC molecule (class I).

Specific immune responses are classified into two different types, based on the components of immune system that mediate the response.

- *Humoral response*, where B cells recognize antigens and produce antibodies that attack them. All these processes are mediated by Th cells. This is the main defense against extra-cellular microbes.

- *Cellular response*, mainly mediated by CTLs which recognize infected cells and eliminate them (help is also needed from Th cells). This response is effective against intra-cellular viruses.

2.1 Tolerance and immune network hypothesis

As we have mentioned before one fundamental feature of the immune system is its capability to recognize self molecules from foreign ones. The main mechanism to induce tolerance to self is the *clonal deletion* of T lymphocytes during intrathymic maturation.

T cells are initially produced with surface receptors (TCR) able to recognize any kind of antigenic determinants (as B receptors), but they undergo a double selection process in the thymus before populating peripheral tissues.

1. *Positive selection*; in this process T cell clones that express receptors able to bind MHC self molecules are permitted to survive, while all the others die. In this step all T cells unable to recognize the complex antigen-self MHC are eliminated and only useful T clones survive.

2. *Negative selection*; T clones that can bind MHC-self peptides complex are eliminated.

At the end of this process of double selection (*thymus selection*), only T clones able to bind self MHC (*MHC restriction*) but unable to react with self molecules can survive.

Despite of this selection process, due to the extension of immune repertorie, lymphocytes can recognize produced antibodies as antigens. Therefore, following an antigenic stimulation, antibodies against the antigen are produced (Ab1), then (Ab2) are produced against Ab1 and so on. Ab2 can be the so called "internal image" of the antigen, that is it could have a binding site similar to the antigen one, since both are able to be bound by the same kind of antigen Ab1, in the same way Ab3 can be similar to Ab1, etc... . This set of reactions suggested to Jerne[2] the idea that the whole immune system be highly connected and it forms an *idiotypic network* (idiotype is the part of antibody that can be bound as an antigenic determinant).

There are many experimental evidences of the existence of the idiotypic network, but it is not yet clear which is its role in the regulation of the immune response; if it is a direct consequence of the huge diversity of immune repertorie or one of the main mechanism of regulation of immune response. In any case, as we will see in the next section, Jerne's hypothesis is taken in consideration in many models.

3   Immune system models

The panorama of immune system models is quite large. In the following we will show some models, this is not a complete review (for a more complete review of immunological models see Perelson and Wisbuch (1997)[3] and Lippert and Behn (1997)[5]) but we will just give a few examples of the possibilty of mathematical modeling in immunology. In the next sections first we will introduce a model to predict the number of parameters necessary to describe the diversity of the molecules involved in the immune response, then we will examine some models based on Jerne's network hypothesis, and finally we will present a "complete" model of humoral response proposed by F. Celada and P.E. Seiden.

3.1   Shape space

One of the main problem in immune system modeling is to threat the huge variety of objects, due to the presence of specific surface receptors on lymphocytes. Perelson and Oster[6] introduced the idea of *shape space*. The binding between receptor and ligand, that characterizes the specifity of response, is due to different kinds of forces, i.e. eletrostatic, Van der Waals, hydrogen bonds, ... but we should also consider other aspects such as the geometry of molecules involved, their sizes and shapes, their charge, etc.. .

Let us suppose that all these features can be modelled using a set of N parameters. A *cognate recognition* (a chemical bond) between a receptor and a ligand will be possible if they assume complementary values of these parameters

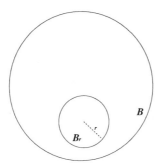

Figure 1: An antibody has a range of interaction of r in the shape space B. It can recognize all the idiotypes included in hypersphere denoted by B_r.

(opposite charges, complementary shapes, etc..) . This set of parameters is called *generalized shape*, the space of parameters S *shape space* and it is a N-dimensional space. The metric of S is not necessarily the euclidean one, but it depends on the choice of parameters.

Following Perelson and Oster, let us to estimate the number of different antibodies recquired to recognize any possible antigen. We shall assume that antibody shapes are randomly distributed throughout a finite volume B with uniform density. This is a strong assumption partly justified by our ignorance about the actual distribution. But this distribution is not surely random, in fact there are some "holes" (i.e. some kinds of antibodies are absent or, at least, inactive) since the immune system cannot recognize self molecules, furthermore the evolutive pressure has driven the system to develop more effective response against some pathogens increasing the density of shapes in correspondence of the more frequent antigenic determinants (we can release this hypothesis using a weaker one, with the same qualititave results, see Perelson and Oster (1979) [6]). Let us also assume that an antibody could recognize all the antigens within a distance r , that is it can bind all the antigens inside the N-dimensional ball centered in its position in the shape space B_r (see Fig. 1). Therefore the corrispondence between antigen-antibody is not 1:1, as the so called key-lock model suggests, and a finite set of antibodies can recognize an actually infinite set of different antigens. If we consider B as the N-dimensional ball and the immune repetorie made of n different non-overlapping shapes of antibodies, the probability p that an antigen is not recognized is:

$$p = (1 - \frac{B_r}{B})^n \simeq e^{-nB_r/B}. \qquad (1)$$

If we want a "good" immune system, its probability of recognizing any kind of antigen should be close to 1, then p close to zero. The frequency B or T cells bind an epitope has been measured[4] to be of the order 10^{-5}. Using this value for B_r/B ratio in eq.(1), if $n = 10^5$, then $p = 1/e \simeq 37\%$ of antigenic determinants can be recognized and the system doesn't work properly, for $n = 10^6$ we get $p = 4.5 \times 10^{-5}$, i.e. every antigen can be bound.

This estimate is an a good agreement with experimental data, in fact the smallest known immune system is that of tadpole, it has order 10^6 lymphocytes and thus the size of its repertorie is $10^5 - 10^6$. Smaller repertories have not been discovered yet, but for value of $n \simeq 10^4$ or less this immune system would be ineffective.

3.2 Immune network models

One of the main features of the immune system is the existence of a memory mechanism. The proposed explanations are of two different kinds:

- **static explanation**: Memory is due to the proliferation, after the first antigen recognition, of specific B and T clones with long lifetime. Another possible mechanism is that the antigen remains trapped in the organism bound on the surface of dendritic cells, and this periodically re-stimulates B and T cells.

- **dynamical explanation.** The presence of the idiotypic network proposed by N.Jerne, as described in section (2.1), is responsible for memory. The idea is the following: for each kind of antibodies there are two stable states; one characterized by a low (or null) concentration of a certain antibody (or B cell) Ab1, and in the other high value of Ab1 concentration is present. In this last case the system can react quickly and strongly when an antigen (with its receptors complementary to Ab1) is introduced. The transition between the two states is triggered by the first antigenic stimulation. Due to the huge diversity of the immune repertorie and of the possible antigens the number of these kind of stable fixed points should be very large. Therefore the different kinds of antibodies are coupled together, and the fixed points should be fixed points of the whole network. The presence of a certain antigen stimulates Ab1 antibodies, these stimulate their complementary Ab2 antibodies (that could be similar to the antigen, *internal image* of the antigen), and so on. Of course if all these interactions are strictly positive we have a percolation along all the network and the system does not work properly.

Most of the models assumes the idiotypic network as the main mechanism of regulation;[5,7] for such kinds of models it is possible to use most of mathe-

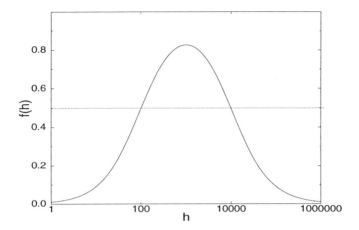

Figure 2: Log-bell shape activation function $f(h)$ for $p = 1$, $\theta_1 = 100$ and $\theta_2 = 10000$.

matical tecniques studied in the framework of neural networks and spin glasses theory.

Usually a lattice is introduced to represent the shape space, we will label its points by the index i. Each point of the lattice takes the value of the concentration of antibodies (or B cells) x_i, that have their receptors described by the parameters that identifies the i site. These concentrations can assume continuous values (differential equation and coupled map models) or discrete (cellular automata). The interactions between the sites depend on the model we consider.

As an example we will consider here the model proposed by Weisbuch, De Boer and Perelson.[8] In the following according to Perelson and Weisbuch (1997)[3] we will call it B model, because it refers to B cells. Different variations of this model can be found in De Boer (1988),[9] De Boe et al. (1989),[10] De Boer and Perelson (1991)[11] and in Stadle et al. (1993).[12]

B model is a set of N differential equations, one for each clone population x_i (i varies from 1 to N, number of points in the shape space).

$$\frac{dx_i}{dt} = s + x_i(f(h_i) - d)$$

where s is the source term due to B cell production by bone marrow, $f(h)$ defines the proliferation rate, d is the death rate, and h_i is due to the interaction of the i-th clone with all the others.

$$h_i = \sum_j J_{ij} x_j \; ,$$

where J_{ij} gives the affinity between clones x_i and x_j. The choice of function f is crucial, experimental evidence and the chemistry of cross-linking suggests a symmetric log-bell shape curve. In the B model case Weisbuch, De Boer and Perelson choose the following function:

$$f(h_i) = p \frac{h_i}{\theta_1 + h_i} \frac{\theta_2}{\theta_2 + h_i}$$

with $\theta_2 \gg \theta_1 > 0$ and $p > d$. In Fig. 2 we plot $f(h_i)$. We can observe that for small and large values (comparered to θ_2 and θ_1 respectively) of h_i we have $f(h_i) < d$ and the network interactions have a *suppressive* function, vice versa in the middle of the curve (i.e. approximatively between the threshold θ_1 and θ_2 if $d \simeq p/2$) $f(h_i) > d$ and the network has a *stimulatory* function.

The topology of the network is definied by the values of J_{ij}. We have a limited knowledge of these values from experiments, a simple choice is to the Cayley tree architecture (another possible choice is using random values, see Lefevre and Parisi (1993) [13]). The choice of the Cayley tree with m connections per site (see Fig. 3) is justified by the necessity to use an not completely connected interaction matrix J_{ij} (because idiotypes can interact just with few others) and, *a posteriori*, by the results, in particular the presence of localized attractors. Then, in the presence of an antigenic stimulation with idiotype 1, we get:

$$h_1 = mx_2 + \zeta a$$
$$h_2 = x_1 + (m-1)x_3$$
$$h_3 = x_2 + (m-1)x_4$$
$$\cdots$$
$$h_i = x_{i-1} + (m-1)x_{i+1}$$

where m is the number of connections per site, a is the concentration of the antigen and ζ is antigen-antibody affinity.

This system has a set of localized attractors. The learning mechanism following to an antigenic stimulation (immune memory) is local and it depends on the variation of the configurations of attractors. Therefore it is very different from neural network models where, usually, attractors are delocalized and the memory is due to the variation of synaptic weights J_{ij}.

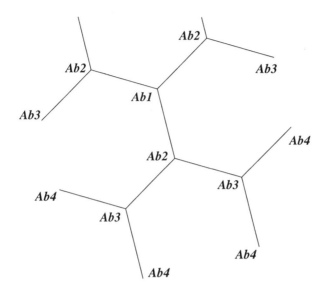

Figure 3: Cayley tree architecture for $m = 3$ connections per site. Antigens interact with the root of tree (Ab1). Ab1 can also interact with Ab2, Ab2 with Ab1 and Ab3, etc..

There are three kinds of immune memory attractors, *vaccination, tolerance* and *percolation*. In this last case (percolation) all the system is activated and localization is lost.

The system evolves towards to one kind of attractor depending on initial conditions (choices of J_{ij} and antigen dose) and values of parameters.

This model has been extended to study the presentation of two complementary antigens,[14] as an hormon and its recptor, because there are autoimmune diseases characterized by such kind of antigens.

Many immune response models based on cellular automata have been introduced.[15,16,17] We shall not discuss these models here (for a review see Stauffer (1997)[18]), but we shall describe a microscopic model based on a cellular automaton proposed by Celada and Seiden.

4 Celada-Seiden model for humoral response

A cellular automaton model for humoral response has been proposed by Celada and Seiden (to be called CS herein).[19,20]

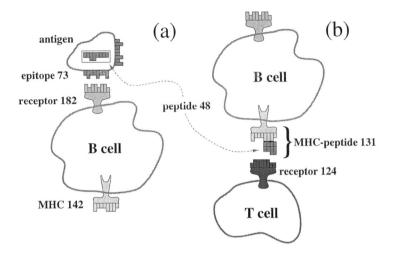

Figure 4: Entities in the Celada-Seiden model simulation for the case of 8 bits. (a) Process of recognition, internalization and (b) presentation of antigen by a B cell to a T cell. Numbers indicate decimal values of the binary strings.

The CS automaton modifies and extends usual cellular automaton rules,[21] allowing probabilistic evolution, making the evolution rules depend only on entities on the same site and permitting entities to move to neighboring sites. These are typical features of reactive lattice gases (see Boo *et al.* (1996)[22] for a review). The CS automaton is defined as a two dimensional lattice, usually of a small size (15 × 15), that represents a small part of the body.

4.1 Components of the model system

An important characteristic of the CS model is the simulation of diversity in antigens and response by the introduction of several clonotypic elements (e.g., epitopes, peptides, and receptors) represented by binary strings. The objects present in the model are:

- cells (APC, B cells and T cells);

- antibodies;

- antigens.

B and T cells are represented by different clones, each clone is characterized by its surface receptor which is modeled by a binary string of N bits (the use of bitstrings to model receptors was introduced by Farmer *et al.* [23]) with a fixed directional reading frame. In Fig. 4a we show how these objects are represented in the model. Each clonotypic set of cells has a diversity of 2^N; for the simulations described in this paper we use $N = 8$, so we have a diversity of 256 (in comparison to $\sim 10^9$ of the real system in man), also the antigens have the same order of diversity in our simulations, therefore we have a complete repertorie. The number of possible states in a site is therefore very high, for example if the maximum number of cells of each kind is 1000, the number of states in each site is $\sim 10^{1536}$.

Besides the receptor (BCR), B cells have MHC class II molecules on their surface. In the model they are also represented by a binary string of N bits. MHC's are involved in the process of B-cell activation. MHC diversity in a given body is very small, less than 10 different MHC molecules are present. We have generally used one or two different kinds of MHC molecules in the simulations.

We have used a complete repertoire for B cells, i.e., all can be produced by the system and, for the parameters we use here, the average occurrence of each clonotype in the starting population is approximately one. For T cells we begin with the complete repertoire, but they are filtered by the *thymus* before putting them in the lattice. The thymus is an organ through which T cells must pass before they can mature. In the thymus they are exposed to self-antigens presented on APCs. If they respond too strongly or not at all they are killed. Therefore, the thymus acts as a filter to remove dangerous or non-responsive T cells. It is the body's first line of defense against autoimmune disease. Including a thymus, results in a T-cell population that has restricted diversity.

APCs represent non-specific antigen presenting cells (such as macrophages), they have no specific receptor (they can bind any type of antigen with a fixed probability) but they have the same MHC as B cells.

Antigens (Ag) are made of two different parts: *epitopes* and *peptides*. The epitope is the portion of an antigen that can be bound by the BCR; after this event the peptide (or better a peptide of the antigen, because generally an antigen has more than one peptide or epitope) is presented on an MHC molecule for T-cell cognate recognition (we show a sketch of the process in Fig. 4b).

Antibodies (Ab) are also made of two parts. They have a receptor (*paratope*), that is represented by the same string as the BCR of the parent B cell which produced them and, optionally, a peptide (*idiopeptide*).

4.2 Interactions

There are precise interaction rules; the allowed interactions are mainly of two types:

- *specific*, as between antigen and antibody, antigen and BCR, or MHC-peptide complex and T cell receptor; this kind of interaction has a probability of interaction (affinity) evaluated according to the number of complementary bits between the binary strings that represent the receptors. In our simulations we have used a probability of 1 in case of perfect complementary, 0.05 for the case of one mismatching bit, and 0 for more than one mismatching bit.

- *non-specific* as APC-antigen interaction. The interaction takes place with a fixed clone-independent probability, typically equal to 0.002.

The main process of humoral response is B- and T-cell proliferation (clonal growth) and consequent antibody production. The process is divided into four parts: antigen-BCR interaction, endocytosis and peptide presentation on MHC, T cell recognition of the MHC/peptide complex, and finally, cell proliferation and differentiation into plasma cells (which produce antibodies) and memory cells (which provide the possibility of a further enhanced response). For each antigen several B-cell clones have receptors able to bind it. After binding, the B cell processes the antigen and presents it on its MHC and then waits for a recognition/binding event by a T cell. This is the signal for the B cells to proceed in the response. The signal may fail to be delivered either because the specific T cells have been negatively selected in the thymus, or because the probability of finding the right combination of B and T cells in the same site is low. The latter situation may change with time, e.g., after the T cells have proliferated through stimulation by APC presenting the same antigen. After they have received this second signal B cells can divide, producing memory and plasma cells. The plasma cells secrete antibodies, having the same receptor as the B cell, which can bind the antigen just as the BCR does. Upon T-B cell interaction the T cells also proliferate.

4.3 Simulations

During a time step each site is considered individually. Each entity in a site is given an opportunity to take part in all interactions for which it is able. The success or failure of an interaction is determined by comparison of its probability with a random number. Although all possible interactions are considered, an entity can have at most one successful interaction on any one time step.

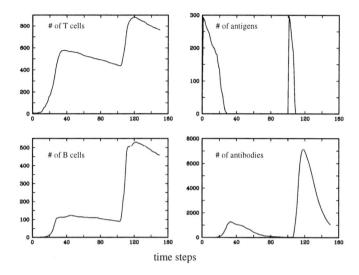

Figure 5: A typical immunization experiment.

After the interactions are determined, the entities are allowed to die (with some half life), stimulated cells divide, new cells are born, and antibodies are generated. Finally, the entities are given an opportunity to diffuse to neighboring sites. This constitutes a time step and the entire process is repeated for as many time steps as desired.

To give a rough idea of the functioning of the CS model we show a typical simulation of *immunization* in Fig. 5. The system initially has no antigen, no antibodies, and 1000 B cells, 1000 T cells and 1000 APCs uniformly distributed in space and receptor type (except that the T cells have been processed in the thymus); no interaction among species is present and the system is in a steady state where the natural death rate equals the birth rate. We start by injecting a single type of antigen (same epitopes and peptides). The antigens are at first bound primarily by the APCs since, although weakly binding, they are much more plentiful than the rare B cells that match the antigen. T cells will stimulate these APCs and then divide to form populous clones. There will then be sufficient antigen responding T cells to easily find the few responding B cells. Upon stimulation from the T cells the B cells also divide and form significant clones. Finally, antibodies are produced and the antigen is removed. This is the *primary response*.

If later more of the same antigen is injected (e.g., at time step 100 of

Fig. 5) it is removed much more rapidly because the system has an appreciable population of B and T memory cells induced by the primary antigen dose. This is the *secondary response* and it can be so strong and swift that the antigen is eliminated before it can do any damage.

A number of other aspects of the humoral immune system have been studied with this model, e.g., response to various levels of antigen dose [19], hypermutation and affinity maturation [24], and thymus function. [25]

4.4 Infection

In this section we shall discuss an extension of the basic Celada-Seiden model for humoral response. The aim is to study the effects of a viral infection, this is the first step toward a more complete implementation of cellular response; for a more complete description see Bezz *et al.* (1995) [26] and (1997). [27]

A first set of simulations has been performed just considering an antigen with an infective function, i.e., capability to penetrate cells and to multiply inside them, like a virus or an intracellular parasite. It has a given probability P_i per time step to infect any B cell or APC present in the same site. The target cells are the cells of immune system: B cells and APC. The choice to infect only these limits the cell species in the model but infections of this type are known, for example, *Epstein-Barr* virus which infects B cells and *Leishmania major* an intracellular parasite of macrophages.

Infected cells continue their *normal* life while the virus duplicates inside them with a constant growth rate (r) (this may be a drastic approximation of realistic situations). That is the number of viruses inside the cell at time n (V_I^n) grows according to:

$$V_I^{n+1} = r \times V_I^n$$

When the number of viruses inside the cell exceeds a fixed threshold (V_{max}), the cell is destroyed and V_{max} viruses are freed. The virus is shielded from antibodies when it is inside the cell but can be destroyed when it is outside. In all simulations described in this section, featuring only humoral response, virus inside the cells is safe.

Simulations was performed by introducing a single injection of virus and then observing the response of the system as a function of V_{max} and P_i.

Three different final states are present:

- indefinite growth of the virus (**V**, diseased state), see Fig. 6 (right-hand panels);

- elimination of the virus (**IS**, immune state), see Fig. 6 (left-hand panels);

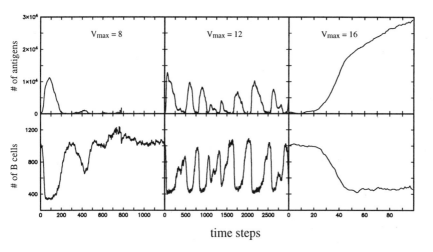

time steps

Figure 6: Evolution of the population of B cells and antigens for three cases of V_{max} with $P_i = 0.05$.

- an oscillatory state (**O**), see Fig. 6 (middle panels).

Fig. 6 presents the total number of B cells and virus versus time. There are two fixed point type basins of attraction **IS** and **V**. The **O** state appears at the border between these two.

If we look at the antigen concentration in the **O** case we see low numbers of B cells correspond to high values of virus, i.e., they are in phase opposition

The occurrence of these global oscillations depends on the diffusion of cells, antibodies and virus throughout the whole body array. Increasing the diffusion constant causes the oscillations to become more regular, fluctuations in time delay between high virus concentration phases are smaller, i.e., frequency modulation decreases. For low values of the diffusion constant the system is less synchronized and at a certain threshold value global oscillations disappear.

These three final states are also present if T killer (Tk) cells and a cross-regulation mechanism between Tk and B cell activation are added in the system. Tk cells can recognize and kill infected cells with a fixed probability (they are not specific here while in reality are). Moreover a cross-regulation mechanism between humoral response and Tk response is introduced: the activation of Tk cells triggers the emission of one kind of cytokines that can inhibit B cell activation, vice versa B cell activation (B - T helper cell interaction) produces another kind of citokines inhibiting Tk activation. The presence of Tk cells obviously improves immune response and, through cross-regulation, biases the response towards on of the two branches (antibodies or Tk cells mediated), as is observed *in vivo*.

So upon introducing an infection mechanism two fixed points are obtained: one characterized by indefinite virus growth (**V**), the other by elimination of the virus (**IS**); these states can be identified with *disease* and with *recovery* from a viral infection. In the case of evolution to **V**, our simulations show a condition of *immunodeficiency*, i.e., a sudden reduction of the number of cells of the immune system. Cases of this sort of behavior aren't known for B cells, but there are a lot of data for T helper cells in AIDS (*acquired immunodeficiency syndrome*). The mechanism is different because HIV (the virus of AIDS) infects and kills T cells.

In this model an oscillating mode **O** is also present. Oscillatory immune responses are found in various experiments *in vivo* and *in vitro*.[28,29,30]

In this system the feedback mechanism is strictly connected to viral growth. Oscillations are found in some infectious diseases. Lo *et al.*[31] have studied an infection due to *Mycoplasma fermentans* isolated from Kaposi's sarcoma of a patient with AIDS. Another kind of infectious disease where an oscillatory pattern can be found is malaria. Malaria is mainly due to two parasites *Plasmodium vivax* and *Plasmodium falciparum* and it presents periodic sharp episodes of high fever (*paroxysms*).[32]

5 Conclusions

Immune system is a widely studied system, a lot of processes are well known also from a molecular point of view, and the technical skills to modify most of them are currently avalaible. Nevertheless many crucial regulation mechanisms are not fully understood yet, such as how really immune memory works (idiotypic network vs. memory cells hypothesis) or self tolerance. The introduction of theoretical models could help understanding which are the minimal mechanisms able to reproduce some experimental results.

Simulations (*in machina* experiments[19]) cannot, of course, replace *in vivo* and *in vitro* experiments but they could be an useful, and relatively cheap tool in the planning of the more expensive real biological experiments.

References

1. A.K. Abbas, A.H. Lichtman and J.S. Pober, *Cellular and Molecular Immunology*, (Saunders, 1991).
2. N. K. Jerne, Ann. Immunol. (Inst. Pasteur), **125 C**, 373 (1974).
3. A.S. Perelson and G. Weisbuch to appear in Reviews of Modern Physics.

4. N.R. Klinman and J.L. Press, Transplant Rev., **24**, 41 (1975).
5. K. Lippert and U. Behn, *Annual Reviews of Computational Physics* vol. V, (World Scientific, Singapore 1997).
6. A.S. Perelson and G.F. Oster, J. Theor. Biol. **81**, 645 (1979).
7. Immunological Review **110**, (1989)
8. G. Weisbuch, R. De Boer and A.S. Perelson, J. Theor. Biol., **146**, 483 (1990).
9. R.J. De Boer. In *Theoretical Immunology, Part Two*, A.S. Perelson, editor (Addison-Wesley, Redwood City, Ca. 1988).
10. R.J. De Boer and P. Hogeweg, Bull. Math Biol. **51**, 223 (1989).
11. R.J. De Boer and A.S. Perelson, J. Theor. Biol. **149**, 381 (1991).
12. P.F. Stadler, P.Schusterand A.S. Perelson, submitted to Bull. Math Biol. (1993).
13. O. Lefévre and G.Parisi, Network **4**, 39-65 (1993).
14. G. Weisbuch, R.M. Zorzenon dos Santos and A. Neumann, J. Theor. Biol. **163**, 237 (1993).
15. M. Kaufman, J. Urbain and R. Thomas, J. Theor. Biol. **114**, 527 (1985).
16. R. De Boer, J.D. Van der Laaan and P. Hogweg, in *Thinking about Biology*, F.J. Varela and W.D. Stein, editors (Addison-Wesley, 1992).
17. D. Stauffer and G. Weisbuch, Physica A **180**, 42 (1992)
18. D. Stauffer *Sex, Money, War, and Computers*, Springer Lecture Notes in Computational Science and Engineering, in press (Springer, Berlin 1997).
19. F. Celada and P.E. Seiden, J. Theor. Biol. **158**, 329 (1992).
20. F. Celada and P.E. Seiden, Immunol. Today **13**, 56 (1992).
21. S. Wolfram, *Theory and Applications of Cellular Automata*, (World Scientific, Singapore 1986).
22. J.P. Boon, R. Dab, R. Kapral and A. Lawniczak, Phys. Rep. **273**, 55 (1996).
23. J.D. Farmer, N.H. Packard and A.S. Perelson, Physica **22D**, 187-204 (1986)
24. F. Celada and P.E. Seiden, Eur. J. Immunol. **26**, 1350 (1996).
25. D. Morpurgo, R. Serenthà, P.E. Seiden and F. Celada, Int. Immunol. **7**, 4, 505 (1995).
26. M.Bezzi, S.Ruffo and P. Seiden, in *Cellular automata for cellular response*, Conference Proceedings Vol.48 "National Workshop on Nonlinear Dynamics", M.Costato, A.Degasperis and M. Milani, editors (SIF, Bologna, 1995).
27. M.Bezzi, F.Celada, S.Ruffo and P. Seiden, Physica A **245/1-2**, 145-163 (1997).
28. S. Britton and G. Moller, J. Immunol. **100**, 1326 (1968).

29. J. Hiernaux, P. Baker and C. Delisi, *et al.*, J. Immunol. **128**, 1054 (1982).

30. D.A. Lawrence, Cellular Immunology **49**, 81, (1980).

31. S. Lo and D. Wear, *et al.*, Clinical Infectious Diseases **17 (Suppl 1)** S283-8 (1993).

32. N.D. Karunaweera and G.E. Grau, *et al.*, Proc. Natl. Acad. Sci. USA **89**, 3200 (1982).

DISCRETE MODELS FOR SIMULATING BIOLOGICAL SYSTEMS

PHILIP E. SEIDEN

IBM Research Center, Yorktown Heights, NY 10598

Introduction

n the last 10 to 20 years modelling biological systems by discrete techniques, uch as cellular automata or lattice gases, has become a small but thriving ndustry. In my lectures I first provided a detailed description of one model in lepth, my own immune simulation (IMMSIM). Most of this has been published n Refs. 16 and 3, and in:

- F. Celada and P.E. Seiden, *A Computer Model of Cellular Interactions in the Immune System*, Immunol. Today **13**, 56-62 (1992).

- F. Celada and P.E. Seiden, *Affinity Maturation and Hypermutation in a Simulation of the Humoral Response*, European J. Immunology **26**, 1350-1358 (1996).

The second part of the lectures provided a description of the use of discrete echniques in other areas of biology. A list of some references on the use of liscrete modelling in biology was provided and is attached below. Six of these vere discussed explicitly. They were chosen both because of their breadth and because they could be readily explained in the time available. The specific papers discussed are:

1. Use of discrete automata for describing and understanding the dynamics of a system under a specified set of rules. The example is in immunology but the technique is useful in many areas of biology. (Ref. 20)

2. Ventricular fibrillation - the study of the propagation of electrical signals in the heart. (Refs. 18 and 7)

3. Tumor growth - propagation and treatment of tumors. (Ref. 25)

4. Cell migration - dynamics of migrating cells. (Ref. 1)

5. Virulence of pathogens - competence and transmission of pathogens in competing for hosts. (Ref. 10)

6. Locomotion - legged models of motion. (Ref. 21)

Ref. 11 is of special interest since it provides an interesting review of th use of cellular automata in biology. The rest of the list should be explored t see the use of discrete dynamics in other, and often more complex, problems

2 Some References on the Use of Discrete Modeling in Biology

1. Y. Lee, S. Kouvroukoglou, L.V. McIntire and K. Zygourakis, *A cellula automaton model for the proliferation of migrating contact-inhibited cell* Biophys J. **69**, 1284-98 (1995).

A cellular automaton is used to develop a model describing the prolifera tion dynamics of populations of migrating, contact-inhibited cells. Sim ulations are carried out on two-dimensional networks of computationa sites that are finite-state automata. The discrete model incorporate all the essential features of the cell locomotion and division processes including the complicated dynamic phenomena occurring when cells col lide. In addition, model parameters can be evaluated by using data from long-term tracking and analysis of cell locomotion. Simulation result are analyzed to determine how the competing processes of contact in hibition and cell migration affect the proliferation rates. The relatio between cell density and contact inhibition is probed by following th temporal evolution of the population-average speed of locomotion. Ou results show that the seeding cell density, the population-average spee of locomotion, and the spatial distribution of the seed cells are crucial pa rameters in determining the temporal evolution of cell proliferation rate The model successfully predicts the effect of cell motility on the growth of isolated megacolonies of keratinocytes, and simulation results agre very well with experimental data. Model predictions also agree well with experimentally measured proliferation rates of bovine pulmonary arter endothelial cells (BPAE) cultured in the presence of a growth facto (bFGF) that up-regulates cell motility.

2. J. Smolle, R. Hofmann-Wellenhof, R. Kofler, L. Cerroni, J. Haas an H. Kerl, *Computer simulations of histologic patterns in melanoma us ing a cellular automaton provide correlations with prognosis*, J. Invest Dermatol. **105**, 797-801 (1995).

Computer simulations have been used frequently in the life sciences t investigate the mechanisms of morphologic pattern formation. The cellu lar automaton program SMN5 is designed to simulate tumor growth an to estimate biologic properties by comparing real tumor patterns with computer-simulated reference patterns. This method was applied to 195 cases of primary melanoma of the skin. S-100-stained sections were evaluated by image analysis and compared statistically to a reference set of 4000 simulated patterns. Estimates of tumor cell proliferation, motility, cell loss, cohesion, stroma destruction, and intercellular signals (autocrine and paracrine factors affecting growth, motility, and cell loss) were calculated. Twelve of 18 estimated parameters correlated significantly with tumor progression, as indicated by vertical tumor thickness (linear regression analysis: $p \leq 0.05$), and 13 of 18 parameters carried prognostic significance (log rank test: $p \leq 0.05$). Poor prognosis was associated particularly with a pronounced increase in the estimates of proliferation, tumor cell motility, and stromal degradation. Poor prognosis was also associated with a decrease in the estimates of cell loss, tumor cell cohesion, and paracrine growth factor dependence. In multivariate analysis using Cox's proportional hazard model, stromal degradation and motility showed prognostic information in addition to conventional prognostic parameters. The study shows that analytical comparison of real tumors with computer-simulated patterns of a cellular automaton facilitates a functional interpretation of tumor morphology, which carries prognostic significance in cutaneous melanoma.

3. D. Morpurgo, R. Serentha, P.E. Seiden and F. Celada, *Modelling thymic functions in a cellular automaton*, Int. Immunol. **7**, 505-16 (1995).

Along the lines developed by Celada and Seiden, for simulating an immune system by means of cellular automata, we have constructed a 'thymus' where T cells undergo positive and negative selection. The populations thus 'matured' have been analyzed and their performance has been tested in machina. The key feature of this thymus is to allow chance meeting and possible interaction between newly born T cells and antigen presenting cells. The latter represent both the epithelial and the dendritic cells of the biological organ and are equipped with MHC molecules that can accommodate selected self peptides. All possible specificities are represented among the virgin T cells entering the thymus, but this diversity is drastically reduced by the time they exit as mature elements. In the model organ the fate of T cells, i.e. whether they will undergo proliferation or apoptosis, is governed by their capacity to recognize MHCs and the affinity of this interaction. Crucial parameters turn out to be the concentration of presenting cells, the number of types of MHC per cell, the 'size of self' in terms of the number of different peptides and their prevalence. According to the results, events in the automaton can realize unforeseen cooperations and competitions among receptors, depending upon the interaction order and frequency, and ultimately determine the rescue or the killing of thymocytes. Thus the making of the mature T repertoire has a random component and cannot be completely predicted.

4. H. Kasmacher-Leidinger and H. Schmid-Schonbein, *Complex dynamic order in ventricular fibrillation*, J. Electrocardiol. **27**, 287-99 (1994).

Self-sustained circus movement of excitation has long been discussed as the underlying mechanism of ventricular fibrillation. This concept now appears to have found general acceptance. Mapping studies are very expensive and do not permit observations to be isolated from varying external influences. The authors have therefore used a simple cellular automation for studying the basic principles of excitation spreading during ventricular fibrillation under various conditions. Highly ordered spiral waves described in other computer models, previously interpreted as fibrillation were encountered. Comparisons of pseudo-electrocardiograms created by the cellular automaton with original electrocardiographic recordings showed similarity in the time series and in fast Fourier transformation. Modulation of refractory times often led to the termination of model fibrillation. Clinical evidence suggests that maintenance of the $1/f$ (RR-variability) variations produced by the autonomic nervous system exerts a protective effect on the evolution of ventricular fibrillation.

5. E. Szathmary, *Toy models for simple forms of multicellularity, soma and germ*, J. Theor. Biol. **169**, 125-32 (1994).

Simple models for the development, maintenance, and origins of primitive, one-or two-dimensional "toy organism" are presented. They are similar to cellular automata with the added combination of internal degrees of freedom that are genetically programmed. The growing rules are implementable by cellular mechanics, biochemistry and genetics. The forms that are dealt with are: sexual unicells, filaments with fragmentation, cell doublets with spore formation, filaments with differentiation of soma and germ, and two-dimensional colonies with early segregation of somatic cells and germ line. The minimum models presented help to understand the feasibility of several important evolutionary transitions. An explanation, supported by the models, for the fact that all three multicellular kingdoms are primarily sexual, is offered by the observation that sexuality in unicells is an excellent preadaptation for development, for the former entails programmed differentiation of cell types (relying in part on the action of homeobox genes), use of cell surface molecules, programmed arrest of cell division, etc. Relevant examples of existing biological systems are also presented.

6. G. Zou and G.N. Phillips Jr., *A cellular automaton model for the regulatory behavior of muscle thin filaments*, J. Biophys. **67**, 11-28 (1994).

Regulation of skeletal muscle contraction is achieved through the interaction of six different proteins: actin, myosin, tropomyosin, and troponins C, I, and T. Many experiments have been performed on the interactions of these proteins, but comparatively less effort has been spent on attempts to integrate the results into a coherent description of the system as a whole. In this paper, we present a new way of approaching the integration problem by using a cellular automaton. We assign rate constants for state changes within each constituent molecule of the muscle thin filament as functions of the states of its neighboring molecules. The automaton shows how the interactions among constituent molecules give rise to the overall regulatory behavior of thin filaments as observed in vitro and is extendable to in vivo measurements. The model is used to predict myosin binding and ATPase activity, and the result is compared with various experimental data. Two important aspects of regulation are revealed by the requirement that the model fit the experimental data: (1) strong interactions must exist between two successively bound myosin heads, and (2) the cooperative binding of calcium to the thin filament can be attributed in a simple way to the interaction between neighboring troponin-tropomyosin units.

7. M.R. Wilby, M. J. Lab and A.V. Holden, *Dynamical model of signal propagation in the heart*, J. Theor. Biol. **168**, 399-406 (1994).

We present a simple model, describing the propagation of a travelling signal in a media that is intended to model phenomena in cardiac tissue. The model is a caricature designed to examine some characteristics of a propagating medium that can support spontaneous and non-stimulated fibrillation behaviour. The model consists of a large number of weakly interacting components, each of which operates according to a simplified set of rules. Specifically it is a form of cellular automaton, with the introduction of probabilistic nature in terms of a random refractory period.

8. D. Stauffer and M. Sahimi, *High-dimensional simulation of simple immunological models*, J. Theor. Biol. **166**, 289-97 (1994).

We propose a simple model for idiotypic-antiidiotypic immunological net works which represents a simplification of the model originally suggeste by Stewart & Varela, and de Boer, van der Laan & Hogeweg. Windov cellular automata enlarge the antibody concentration if the influence c the immediate neighborhood lies between 70 and 99% of its maximur value; in addition random recruitment is possible. We simulate the mode on large square lattices as well as in higher dimensions. The results are i qualitative agreement with those of the earlier model. Moreover, in fiv to ten dimensions we find phase separation for not too large recruitment which then leads to domains growing to "infinity" for infinite times.

9. A. Brass,R.K. Grencis and K.J. Else,*A cellular automata model for helpe T cell subset polarization in chronic and acute infection*, J. Theor. Bio: **166**, 189-200 (1994).

A cellular automata (CA) model has been built to study the interac tion between T-helper subset cells in a secondary lymphoid organ durin, chronic and acute infection. The TH subset cells interacted via shor range cytokine-like factors, each cell type producing an autocrine facto and another factor which suppressed the development and proliferatio: of the other TH cell type. A cell death term was also included suc that T cells not restimulated by antigen within a certain time died to b replaced with new naive cells. The important parameters in the mode were the antigen density entering the lymph node and the propensity o the antigens to induce naive T cells down a specific TH subset pathwaу Many features of the response of the CA were found to match those see: in infections known to induce TH subset polarization. For example, i could be seen that TH cell subset polarization arose as a natural conse quence of the dynamic competition between TH1 and TH2 cytokines t(induce or suppress proliferation and was driven by the antigen produce(by the pathogen.

10. A.E. Kiszewski and A. Spielman, *Virulence of vector-borne pathogens. / stochastic automata model of perpetuation*, Ann. N.Y. Acad. Sci. **740** 249-59 (1994).

To determine how virulence may be perpetuated in populations of vector borne pathogens, we simulated their fitness in a stochastic simulatio: based on cellular automata. Thereby, directly transmissible pathogen: that differed in virulence were permitted to compete for hosts with sim ilarly virulent pathogens that could infect hosts remotely because the⁊ were vector-borne. Fitness was defined as the proportion of the hos population infected with each pathogen at equilibrium. Virulent, directly transmitted pathogens prevailed solely when their infectivity was transient. When duration of infectivity exceeded that of host survival, the less virulent pathogen invariably prevailed. Although remotely transmitted virulent pathogens persisted somewhat longer than did virulent pathogens that were transmitted directly, they never perpetuated themselves. We conclude that populations of vector-borne pathogens may retain pathogenicity somewhat longer than do those that are directly transmitted, but that both kinds of pathogens tend to become nonvirulent.

11. G.B. Ermentrout and L. Edelstein-Keshet, *Cellular automata approaches to biological modeling*, J. Theor. Biol. **160**, 97-133 (1993).

We review a number of biologically motivated cellular automata (CA) that arise in models of excitable and oscillatory media, in developmental biology, in neurobiology, and in population biology. We suggest technical and theoretical arguments that permit greater speed and enhanced realism, and apply these to several classical examples of pattern formation. We also describe CA that arise in models for fibroblast aggregation, branching networks, trail following, and neuronal maps.

12. A. Deutsch, A. Dress and L. Rensing, *Formation of morphological differentiation patterns in the ascomycete Neurospora crassa*, Mech. Dev. **44**, 17-31 (1993).

Morphological differentiation patterns–among them concentric rings and radial zonations–can be induced in the band-mutant of Neurospora crassa by appropriate experimental conditions, in particular by a mere shift of certain salt concentrations in the medium. The role of initial experimental conditions is examined and, furthermore, the influences of artificially induced phase differences are analyzed with respect to pattern formation. While the concentric ring pattern is due to some (endogenous) circadian rhythmicity within every hypha, nothing is known about the underlying mechanism of radial zonation development. Various hypotheses were tested with the help of a cellular automaton model which mimics growth, branching and differentiation of a fungal mycelium. In particular, sufficient conditions are provided which imply the formation of radial spore zonations. These conditions postulate a rather homogeneous microscopic hyphal branching pattern and induction of spore differentiation by means of an activator-inhibitor system. Furthermore, a working hypothesis for the formation of spore patterns in Neurospora crassa is suggested which is based on an extracellular control of fungal differentiation.

13. F.A. Bignone, *Cells-gene interactions simulation on a coupled map la:
tice*, J. Theor. Biol. **161**, 231-49 (1993).

A discontinuous mapping, developed to model gene expression insid
cells and cellular interactions on a lattice, is described. Gene dynam
ics (considered here only as genetic trans regulations) are formulated i
a way similar to the Boolean-neural network (BN) approach through
step function, and gene products are supposed to diffuse on a regular dis
crete lattice of cells, giving rise to nearest neighbour–cellular automat
(CA) type–interactions. The time evolution of the model is discrete an
synchronous. Despite its simple form the system shows a variegate pat
tern of behaviour, and certain regions in parameter space show a ric.
series of bifurcations and multistability already in the case of network
with two genes. The patterns of distribution of the products on th
lattice are similar to the ring dynamics observed in CA following Wol
fram's formulation and two-dimensional (torus) dynamics described i
the Greenberg-Hasting model for excitable media.

14. P.A. Dufort and C.J. Lumsden, *Cellular automaton model of the acti:
cytoskeleton*, Cell. Motil. Cytoskeleton **25**, 87-104 (1993).

We describe a cellular automaton model of the actin cytoskeleton. Th
model incorporates spatial and temporal behavior at the macromolecula
level and is relevant to the viscous nonequilibrium conditions suspectec
to occur in vivo. The model includes cation and nucleotide bindin;
to actin monomers, actin nucleation and polymerization into filaments
cross-linking with alpha-actinin, monomer sequestration with profilin
filament severing, capping and nucleation with gelsolin, binding of pro
filin and gelsolin to membrane-bound phosphatidylinositide biphosphate
(PIP2), and regulation of cross-linking and severing by changing cal
cium levels. We derive 1) equations for the molecular translation and
rotation probabilities required for the cellular automaton simulation ir
terms of molecular size, shape, cytoplasmic viscosity, and temperature
and 2) equations for the binding probabilities of adjacent molecules ir
terms of experimentally determined reaction rate constants. The mode
accurately captures the known characteristics of actin polymerizatior
and subsequent ATP hydrolysis under different cation and nucleotide
conditions. An examination of gelation and sol-gel transitions resulting
from calcium regulation of alpha-actinin and gelsolin predicts an inho-
mogeneous distribution of bound alpha-actinin and F-actin. The double-
bound alpha-actinin (both ends bound to F-actin) is tightly bunched
while single-bound alpha-actinin is moderately bunched and unbound

alpha-actinin is homogeneously distributed. The spatial organization of the alpha-actinin is quantified using estimates of fractal dimension. The simulation results also suggest that actin/alpha-actinin gels may shift from an isotropic to an amorphous phase after shortening of filaments. The gel-sol transition of the model shows excellent agreement with the present theory of polymer gels. The close correspondence of the model's predictions with previous experimental and theoretical results suggests that the model may be pertinent to better understanding the spatial and temporal properties of complex cytoskeletal processes.

15. A.S. Qi, X. Zheng, C.Y. Du and B.S. An, *A cellular automaton model of cancerous growth*, J. Theor. Biol. **161**, 1-12 (1993).

A cellular automaton model describing immune system surveillance against cancer is furnished. In formulating the model, we have taken into account the microscopic mechanisms of cancerous growth, such as the proliferation of cancer cells, the cytotoxic behaviors of the immune system, the mechanical pressure inside the tumor and so forth. The model may describe the Gompertz growth of a cancer. The results are in agreement with experimental observations. The influences of the proliferation rate of cancer cells, the cytotoxic rate and other relevant factors affecting the Gompertz growth are studied.

16. P.E. Seiden and F. Celada, *A model for simulating cognate recognition and response in the immune system*, J. Theor. Biol. **158**, 329-57 (1992).

We have constructed a model of the immune system that focuses on the clonotypic cell types and their interactions with other cells, and with antigens and antibodies. We carry out simulations of the humoral immune system based on a generalized cellular automaton implementation of the model. We propose using computer simulation as a tool for doing experiments in machine, in the computer, as an adjunct to the usual in vivo and in vitro techniques. These experiments would not be intended to replace the usual biological experiments since, in the foreseeable future, a complete enough computer model capable of reliably simulating the whole immune would not be possible. However a model simulating areas of interest could be used for extensively testing ideas to help in the design of the critical biological experiments. Our present model concentrates on the cellular interactions and is quite adept at testing the importance and effects of cellular interactions with other cells, antigens and antibodies. The implementation is quite general and unrestricted allowing most other immune system components to be added with relative ease when desired.

17. M.A. Mainster, *Cellular automata: retinal cells, circulation and patterns* Eye **6**, 420-7 (1992).

Cellular automata modelling is a useful mathematical technique for sim ulating complex biological systems. An area to be studied is broken into a lattice of adjacent cells depicted by picture elements on a computer screen. The initial tissue pattern evolves on the computer screen, di rected by a rule that considers the state of each cell and its neighbours in the lattice. Simulations of wound repair, cell proliferation, retinal cir culation and pigment aggregation serve to illustrate the potential value of cellular automata modelling in ophthalmic research and practice.

18. R.H. Mitchell, A.H. Bailey and J. Anderson, *Cellular automaton model of ventricular fibrillation*, IEEE Trans. Biomed. Eng. **39**, 253-9 (1992).

A theoretical analysis of ventricular fibrillation and the requirements for fibrillation are performed using a discrete element neighborhood (cellular automaton) model of ventricular conduction. The model is configured as a 2500 element rectangular grid on the surface of a cylinder. It is shown that vulnerability to fibrillation is strongly influenced by excited state duration which primarily determines the nature of the underlying reentry activity. As excited state duration is increased fibrillation changes from "coarse" macroreentrant activity to the more chaotic "fine" fibrillation sustained by multiple wavelets of microreentry. In general, defibrillation is achieved by a stimulus strong enough to depolarize the majority of relative refractory elements. The threshold for defibrillation is increased for the more irregular microreentrant fibrillation.

19. D. Chowdhury, M. Sahimi and D. Stauffer, *A discrete model for immune surveillance, tumor immunity and cancer*, J. Theor. Biol. **152**, 263-70 (1991).

In this paper we propose a model of tumor immunity in terms of discrete automata where each automation describes the concentration of one particular type of cell involved in immune response. In contrast to the earlier models of normal immune response, there is more than one type of cell surface antigen in this model. As a consequence, the tumor can evade destruction through humoral response by changing its identity. However, the tumor can be killed by the killer cells through cell-mediated response unless protected by a high concentration of the suppressor T cells.

20. D. Chowdhury, M. Sahimi and D. Stauffer, *A unified discrete model of immune response*, J. Theor. Biol. **165**, 207-15 (1990).

In this paper we propose a unified model of immune response, in terms of discrete automata describing the concentrations of the cells constituting the immune network. The model of normal immune response proposed by Kaufman, Urbain and Thomas and that of auto-immune response proposed by Weisbuch, Atlan and Cohen are special cases of this unified model. Moreover, this model also describes the immune response in patients infected by the human immunodeficiency virus (HIV), the virus that is known to cause Acquired Immune Deficiency Syndrome (AIDS).

21. R. Tomovic, R. Anastasijevic, J. Vuco and D. Tepavac, *The study of locomotion by finite state models*, Biol. Cybern. **63**, 271-6 (1990).

A methodology to derive finite state models of legged locomotion is outlined. Background data for model derivation are joint angle functions and gait diagrams. The method is used to describe the walking of the cat in terms of an abstract automaton. The main features of finite state descriptions of legged locomotion are described. Such models are presenting locomotion invariants of a species in explicit form. It is emphasized that finite state models provide insight into structural features of motor control organization such as decomposition into subsystems, interaction between centralized and decentralized control, and the role of control levels. The finite state model of locomotion can be helpful in suggesting experiments pertinent to the study of motor control and interpretation of experimental results.

22. P. Hogeweg, *Local T-T cell and T-B cell interactions: a cellular automaton approach*, Immunol. Lett. **22**, 113-22 (1989).

In this paper we use cellular automata to study growth factor (IL-2) dependent proliferation of helper T cell (Th) and B cell clones at the level of individual cells. We argue that such a spatially-and individual-oriented approach can provide important insights, not obtainable by more conventional modelling approaches in which the immune system is modelled as a well mixed collection of clones. Two questions are examined: (1) under which conditions can a cell which produces its own growth factor (i.e. Th cells) be regulated by it; and (2) if a growth factor is effective only locally, and if both Th and B cells depend on growth factors excreted by Th cells, how can the spatial segregation of T cells and B cells in lymphoid organs and/or at acute infection sites be explained? The results show that, firstly, autocrine regulation can indeed occur in two ways: it can ensure (a) that the cell reacts only on its growth factor when packed inside tissue of arbitrary cells or (b), that the cell reacts only when close to other growth factor producing cells; and secondly, segregation of T cells and B cells results automatically from simple assumptions about the interaction and proliferation of the cells, notwithstanding the fact that proliferation is slowed down by this segregation.

23. I.R. Cohen and H. Atlan, *Network regulation of autoimmunity: an automation model*, J. Autoimmun. **2**, 613-25 (1989).

Four classes of regulatory T lymphocytes have been implicated in the control of experimental autoimmune diseases: a pair of helper and suppressor T lymphocytes that recognize the self-antigen (antigen-specific) and a pair of helper and suppressor T lymphocytes that recognize the autoimmune effector lymphocytes (anti-idiotypic). The anti-idiotypic pair of regulators was detected following vaccination against autoimmune disease using autoimmune effector T clones as vaccines. To learn how the anti-idiotypic regulatory lymphocytes might function in concert with the antigen-specific regulatory lymphocytes, we devised a network in which the cell populations could be viewed as interconnected automata. Analysis of this novel network model suggests how self-tolerance may operate, how progressive autoimmune disease may develop, and how T-cell vaccination can control autoimmune disease.

24. S.R. Hameroff, S.A. Smith and R.C. Watt , *Automaton model of dynamic organization in microtubules*, Ann. N.Y. Acad. Sci. **466**, 949-52 (1986).

25. W. Duchting and T. Vogelsaenger, *Analysis, forecasting, and control of three-dimensional tumor growth and treatment*, J. Med. Syst. **8**, 461-75 (1984).

The main point of this contribution is to show how ideas of control theory, automata theory and computer science can be applied to the field of cancer research. We are stressing the modelling of three-dimensional tumor growth and the simulation of different kinds of tumor therapy (surgery, radiation therapy, chemotherapy). In the future it will be possible to schedule the optimized methods and time of tumor treatment by computer simulation prior to clinical therapy.

MODELING TH1-TH2 REGULATION, ALLERGY, AND HYPOSENSITIZATION

ULRICH BEHN

Institute for Theoretical Physics, University of Leipzig,
Augustusplatz 10, D-04109 Leipzig, Germany.
e-mail:behn@itp.uni-leipzig.de

HOLGER DAMBECK

Institute for Theoretical Physics, University of Leipzig,
Augustusplatz 10, D-04109 Leipzig, Germany.

GERHARD METZNER

Institute for Clinical Immunology and Transfusion Medicine,
Division of Allergy and Clinical Immunology,
University of Leipzig,Härtelstr. 16–18, D-04107 Leipzig, Germany.

1 Introduction

In the recent past it has been revealed that subsets of T-helper cells exist, Th1- and Th2-cells, which differ in their pattern of cytokine production. The characteristic cytokines released by Th1- and Th2-cells have autocatalytic (autocrine) effects on their own phenotype and are mutually inhibitory for the reciprocal phenotype.

It has been demonstrated that in several infectious diseases the T-cell response is polarized, i.e. it is dominated by either Th1- or Th2-cells. For example, the response to extracellular parasites such as helminths is Th2-dominated, whereas the response to intracellular parasites (e.g. certain protozoans) is Th1-dominated.

T-cell regulation seems to be a crucial issue also in allergy. In allergic individuals a Th2-dominated response to allergen has been found instead of a Th1-dominated pattern which is characteristic for non-allergic response. Injection of increasing doses of allergen following an empirically justified schedule of administration is a therapy for type I allergic diseases (hyposensitization or specific immunotherapy). A successful specific immunotherapy is associated with a change from prevalent Th2- to a prevalent Th1-profile of allergen specific T-cells. There are a lot of empirical data, nevertheless the mechanism of hyposensitization still deserves a theoretical understanding. [1]

In this paper we formulate a mathematical model describing T-cell regulation which provides for the first time a theoretical explanation of the Th2-Th1

228

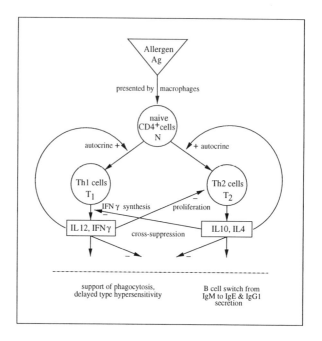

Figure 1: Th1-Th2 regulation via cross-suppressive and autocrine cytokines.

switch due to hyposensitization as a transient dynamical phenomenon.

In the next Section we briefly discuss a (necessarily simplified) scheme of Th1-Th2 regulation mediated by the cytokine network and its relation to allergy underlying our mathematical model.

In Section 3 we formulate the model which describes the population dynamics of the set of allergen-specific naive T-cells, Th1 and Th2-cells, autocrine and cross-suppressive cytokines, and allergen by a system of nonlinear differential equations. We discuss fixed points and the dynamic response to allergen for different parameter settings characterizing both allergic and non-allergic situations.

Section 4 gives a theoretical explanation for the mechanism of hyposensitization in the frame of nonlinear dynamics. It becomes clear why after a successful therapy the desensitized state is only temporary and the therapy should be regularly repeated.

In the concluding Section we briefly comment further perspectives in modeling Th1-Th2 regulation.

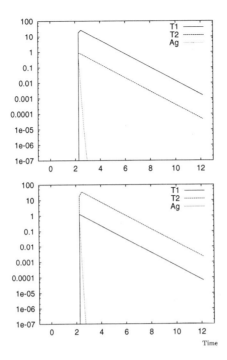

Figure 2: Response of the virgin state to an allergen injection. The figure shows the response of the virgin state $(N, T_1, T_2, Ag) = (\alpha, 0, 0, 0)$ to the injection of an allergen dose $\xi = 1$ for $\mu_2 = 1$ (above) and $\mu_2 = 3$ (below). For values of μ_2 below a critical value μ_2^c ($1.50 < \mu_2^c < 1.55$) the response is Th1-dominated otherwise it is Th2-dominated, i.e. allergic. The critical value depends on the other parameters ($\mu_1 = 1$, $v = 8$, $c = 0.0001$, $\alpha = 10$, $\phi = 0.9$).

2   Th1-Th2 Regulation and Allergy

In this Section we give a simplified description of Th1-Th2 regulation and its relation to allergy which closely follows Powrie and Coffman (1993),[2] cf. Fig. 1.

Allergen (respectively antigen) presented by macrophages trigger naive $CD4^+$ T-cells to differentiate into Th1- and/or Th2-cells. The two subsets of T-helper-cells, Th1 and Th2, differ in the secreted cytokine pattern and in the immune reactions they support. The released cytokines have autocrine but also cross-suppressive effects.[2] Th1-cells secrete IL12 and IFNγ which induce autocrine effects on the production of Th1-cells and act cross-suppressive on the proliferation of Th2-cells, respectively. Th2-cells secrete IL10 and IL4 which have autocrine effects on the production of Th2-cells and cross-suppressive effects on the synthesis of IFNγ.

The reasons why one antigen causes a Th1-dominated and the other one a Th2-dominated immune response are certainly of multiple nature: The presence of cytokines, the type of antigen-presenting cells (APC), the antigen itself, the concentration of the antigen and probably genetic factors determine which subset is dominating the response. [3]

The Th1-subset induces delayed-type hypersensivity reactions and –within a certain limit– the production of other antibody types than IgE. [2,3]

Th2-cells encourage via their cytokines the production of antibodies, particulary IgE-antibodies. [2,3]

Allergy of type I is a typical Th2-response. [3] Specific IgE antibodies are bound to the surface of mast cells or basophils; binding of allergen causes a release of vasoactive mediators and thus promote the allergic cascade. The allergen dose is a major regulator of Th1/Th2-selection. [3] For allergic individuals, medium and high antigen concentration cause a rather Th2-dominated response, whereas only low allergen concentration leads to a Th1-dominated response.

The therapy of hyposensitization by injecting increasing doses of allergen starts with low allergen levels, i.e. provoking a Th1-dominated response. Through Th1-Th2 cross-suppression Th1-dominance may be continued even to higher allergen levels. We hypothesize that after a successful therapy the number of allergen-specific Th2-cells is temporarily reduced because of the dominating Th1-subset. Less Th2-cells means less IgE production, i.e. reduced allergic symptoms.

3   Mathematical Model

3.1   The basic model

The simplified scheme sketched above is the basis of our mathematical model. In the spirit of population dynamics we describe dynamics and interactions of a set of allergen-specific naive T-cells, Th1- and Th2-cells, autocrine and cross-suppressive cytokines, and allergen by a system of nonlinear ordinary differential equations. To reduce the number of variables we do not distinguish the different cytokines secreted by Th1-cells denoting all of them by IF, in a similar way we denote the cytokines secreted by Th2-cells by IL. For better understanding we refrain in the first step from including cytokines coming from other immune processes (later considered as a cytokine background) as well as long-lived memory cells.

Thus we model the nonlinear dynamics of Th1-Th2 regulation in the presence of allergen by the following equations for the mean number of cells, respectively molecules, per unit volume of naive cells ($\hat{N}$), Th1- and Th2-cells

$(\hat{T}_1$ and $\hat{T}_2)$, cytokines IF and IL, and allergen $(\hat{A}g)$:

$$\frac{d\hat{N}}{dt} = -\gamma\,\hat{N} + \rho - \beta_1\,\hat{N}\,\hat{A}g\,IF - \beta_2\,\hat{N}\,\hat{A}g\,IL, \tag{1}$$

$$\frac{d\hat{T}_1}{dt} = -\gamma\,\hat{T}_1 + v\,\beta_1\,\hat{N}\,\hat{A}g\,IF, \tag{2}$$

$$\frac{d\hat{T}_2}{dt} = -\gamma\,\hat{T}_2 + v\,\beta_2\hat{N}\,\hat{A}g\,\frac{IL}{1 + c_1\,IF}, \tag{3}$$

$$\frac{dIF}{dt} = -\delta\,IF + \alpha_1\,\frac{\hat{T}_1}{1 + c_2\,IL}, \tag{4}$$

$$\frac{dIL}{dt} = -\delta\,IL + \alpha_2\,\hat{T}_2, \tag{5}$$

$$\frac{d\hat{A}g}{dt} = \omega(t) - \lambda\,\hat{A}g\,(\hat{T}_1 + \hat{T}_2). \tag{6}$$

Naive cells are produced with a rate ρ and decay as well as the Th1- and Th2-cells with a characteristic time $1/\gamma$. Stimulated naive cells disappear from the pool of naive cells; their number is proportional to $\hat{N} \cdot \hat{A}g \cdot cytokine$, where the last factor mirrors the autocrine effect of the cytocines IF or IL. The parameters β_1 and β_2 allow for differences in the activation of Th1- and Th2-cells by allergen presenting cells. A stimulated naive cell proliferates and produces in the mean v T-cells.

Cytokines decay with a characteristic time $1/\delta$ which is small compared to $1/\gamma$; their production rate is proportional to the number of T-cells. The cross-suppression is modelled by factors of the form $1/(1 + const \cdot cytokine)$ which tend to 1 for low concentrations. Since the cytokines IL secreted from Th2-cells suppress the IF production whereas the cytokines IF secreted from Th1-cells suppress the Th2-proliferation there is an asymmetry in the equations for $\hat{T}_1$ and IF on one side and the equations for $\hat{T}_2$ and IL on the other side.

Allergen is supplied at a rate $\omega(t)$ and is eliminated proportional to the number of Th1- and Th2-cells.

3.2 Adiabatic elimination

Since the life-time of cytokines is short compared to that of T-cells $(1/\delta \ll 1/\gamma)$ cytokines relax fast to a quasi-stationary state dictated by the Th1- and Th2-cells,

$$IL = \frac{\alpha_2}{\delta}\,\hat{T}_2, \tag{7}$$

$$IF = \alpha_1 / (\delta (1 + c_2 \frac{\alpha_2}{\delta} \hat{T}_2)) \hat{T}_1 . \tag{8}$$

This allows to reduce the number of variables by inserting Eqs. (7-8) into (1-3). Measuring the time in units of $1/\gamma$ and rescaling the variables as

$$\hat{N} = \frac{\gamma}{\lambda} N, \quad \hat{T}_1 = \frac{\gamma}{\lambda} T_1, \quad \hat{T}_2 = \frac{\gamma}{\lambda} T_2, \quad \hat{Ag} = \frac{\lambda \delta}{\alpha_1 \beta_1} Ag, \tag{9}$$

leads to the reduced set of equations

$$\frac{dN}{dt} = -N + \alpha - N Ag \frac{T_1}{1 + \mu_2 T_2} - \phi N Ag T_2, \tag{10}$$

$$\frac{dT_1}{dt} = -T_1 + v N Ag \frac{T_1}{1 + \mu_2 T_2}, \tag{11}$$

$$\frac{dT_2}{dt} = -T_2 + v \phi N Ag \frac{T_2}{1 + \mu_1 \dfrac{T_1}{1 + \mu_2 T_2}}, \tag{12}$$

$$\frac{dAg}{dt} = \xi(t) - Ag (T_1 + T_2), \tag{13}$$

where the new parameters $\mu_1, \mu_2, \phi, \alpha$, and $\xi(t)$ are related to the old ones by

$$\mu_1 = \frac{c_1 \alpha_1 \gamma}{\delta \lambda}, \quad \mu_2 = \frac{c_2 \alpha_2 \gamma}{\delta \lambda}, \quad \phi = \frac{\alpha_2 \beta_2}{\alpha_1 \beta_1}, \quad \alpha = \frac{\rho}{\lambda}, \quad \xi(t) = \frac{\alpha_1 \beta_1}{\gamma \delta \lambda} \omega(t). \tag{14}$$

The meaning of the new parameters is obvious: α is the production rate of the naive T-cells and ξ is the injection rate of the allergen in reduced units. μ_1 and μ_2 control the efficiency of the cross-suppression of Th1- and Th2-cells mediated by their cytokines. ϕ regulates the balance of the autocrine effects of the Th1-Th2 system.

3.3 Fixed points

We now discuss the fixed points $\vec{Z} = (N, T_1, T_2, Ag)$ of Eqs. (10-13) in the naive state (i.e. without previous contact to allergen) and for a permanent supply of allergen.

Without allergen, $\xi(t) = 0$, the system has the only fixed point

$$\vec{Z}_0 = (\alpha, 0, 0, 0), \tag{15}$$

which describes a population of naive T-cells maintained by the constant production. Both, Th1- and Th2-populations are absent since the naive cells are not stimulated.

For a constant allergen supply, $\xi(t) = \xi = const$, we have two stable fixed points

$$\vec{Z}_1 = (\frac{\alpha}{1+\xi}, \frac{v\alpha\xi}{1+\xi}, 0, \frac{1+\xi}{v\alpha}), \tag{16}$$

$$\vec{Z}_2 = (\frac{\alpha}{1+\phi\xi}, 0, \frac{v\alpha\phi\xi}{1+\phi\xi}, \frac{1+\phi\xi}{v\alpha\phi}). \tag{17}$$

$\vec{Z}_1$ describes a state of coexistence of naive cells, Th1-cells and allergen, Th-2 cells are extinct. Contrary, $\vec{Z}_2$ describes a state where Th1-cells are extinct. There is a third –though unstable– fixed point $\vec{Z}_3$ where Th1- and Th2-populations coexist.

What is happening if the system is in the naive state described by the fixed point $\vec{Z}_0$ and is exposed for the first time with an allergen? The obvious answer is: nothing, since a stimulation needs *both* allergen and autocrine cytokines. The latter, assumed proportional to Th1- and Th2-cells, are however absent in this state.

3.4 Cytokine background

The discussion of the response to a first allergen contact shows that it is necessary to modify Eqs. (10-13) by taking into account a small *cytokine background* c having its origin in other, comparable, immunological processes. In the spirit of a mean field description it can be assumed to be constant in time and of the same order for both subpopulations. This leads to

$$\frac{dN}{dt} = -N + \alpha - N\,Ag\,(\frac{T_1}{1+\mu_2 T_2} + c) - \phi\,N\,Ag\,(T_2 + c), \tag{18}$$

$$\frac{dT_1}{dt} = -T_1 + v\,N\,Ag\,(\frac{T_1}{1+\mu_2 T_2} + c), \tag{19}$$

$$\frac{dT_2}{dt} = -T_2 + v\,\phi\,N\,Ag\,\frac{T_2 + c}{1 + \mu_1\dfrac{T_1}{1+\mu_2 T_2}}, \tag{20}$$

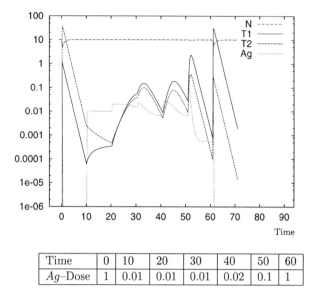

Time	0	10	20	30	40	50	60
Ag–Dose	1	0.01	0.01	0.01	0.02	0.1	1

Figure 3: Successful hyposensitization. After a first encounter of the virgin state with an allergen dose $\xi = 1$ at $t = 0$ which is Th2-dominated (i.e. allergic) follows a sequence of injections in slowly increasing doses (see table). After this treatment the response to the allergen dose $\xi = 1$ given at $t = 60$ is Th1-dominated, i.e. non-allergic. ($\mu_1 = 1, \mu_2 = 2$, $v = 8$, $c = 0.0001$, $\alpha = 10$, $\phi = 0.95$).

$$\frac{dAg}{dt} = \xi(t) - Ag\,(T_1 + T_2). \tag{21}$$

Note, that we neglect the cytokine background in the factors describing suppression because of the smallness of c.

Obviously, we find in the naive state the same fixed point $\vec{Z}_0$ as for (10-13) but now the system allows for a reasonable response to a first exposure with allergen. For a constant allergen supply the system has two stable fixed points to be compared with $\vec{Z}_1$ and $\vec{Z}_2$ in which either Th1- or Th2-cells dominate. Due to the cytokine background the dominated populations are however not extinct. (The correction to Eqs. (16-17) in first order of c can be obtained explicitely.)

3.5 Parameter choice

The parameters are chosen as follows. The dimensionless time is measured in units of the mean life time of T-cells $1/\gamma$. The production rate α of naive cells

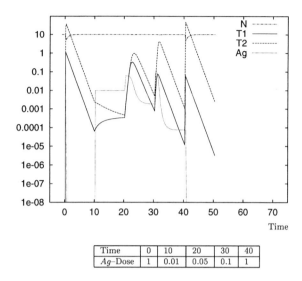

Time	0	10	20	30	40
Ag–Dose	1	0.01	0.05	0.1	1

Figure 4: Unsuccessful treatment. For the same parameters as in Fig. 2 a different schedule of administration of allergen with faster increasing doses is used (see table). The final response to an allergen dose $\xi = 1$ is even stronger Th2-dominated than the primary response.

determining the population of the naive state is chosen such that a unit dose of allergen $\xi = 1$ (e.g. applied by a bee sting) is sufficiently fast eliminated, i.e. in less than a time unit, as it is the case for $\alpha \gtrsim 10$. The proliferation rate of stimulated T-cells v is chosen as 8, the qualitative behaviour of the system is not sensitive to this parameter. The small cytokine background is chosen in most of the presented numerical experiments as $c = 10^{-4}$, with exception of Fig. 7.

The decisive parameters determining the qualitative behaviour of the system are μ_1 and μ_2 measuring the efficiency of the cross-suppression and ϕ controlling the balance of the autocrine effects.

In the allergic state the response to low allergen doses is Th1-dominated in contrast to a Th2-dominance for high doses. For lower doses of allergen autocrine effects alone are important, Th1-cells dominate for $\phi < 1$.

High doses of allergen cause large numbers of T-cells, therefore the numerators $1 + \mu_2 T_2$ and $1 + \mu_1 \dfrac{T_1}{1 + \mu_2 T_2}$ in (18-20) become important. Suppression by Th2-cells is more efficient than suppression by Th1-cells if μ_2 is larger than μ_1. Thus Th1-Th2 regulation is conceived as a competition between

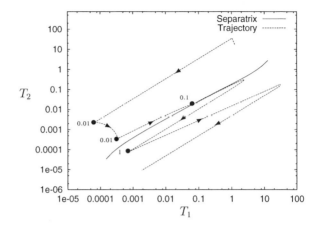

Figure 5: Successful hyposensitization. The figure shows the trajectory for a successful hyposensitization projected onto the $T_1 - T_2$ plane. The instants of injecting a certain dose of allergen are marked by •. A separatrix (solid line) divides the $T_1 - T_2$ plane into two parts in which the response to an allergen dose $\xi = 1$ is either Th2- or Th1-dominated, i.e. allergic respectively non-allergic. The injections make Eqs. (18-21) non-autonomous and allow to cross this separatrix. ($\mu_1 = 1$, $\mu_2 = 3$, $v = 8$, $c = 0.0001$, $\alpha = 10$, $\phi = 0.9$).

autocrine and cross-suppressive effects mediated by the cytokines.

This qualitative reasoning is confirmed by numerical tests. Chosing $\mu_1 = 1$ and $\phi = 0.9$ one finds that for $\mu_2 < \mu_c$ ($\mu_c = 1.50...1.55$) the response of the naive state to a unit dose of allergen is Th1-dominated whereas it is Th2-dominated for $\mu_2 > \mu_c$, cf. Fig. 2. To describe the allergic state we therefore chose in the numerical experiments $\mu_1 = 1$ and $(\phi, \mu_2) = (0.95, 2)$ or $(0.9, 3)$ to ensure a pronounced Th2-dominated response of the naive state to a unit dose of allergen.

4   Hyposensitization

The strategy of hyposensitization –proved to be successful in practice– is explained for the example of a bee venom allergy.

The therapy starts with the injection of a small dose of allergen which is about $10^{-4}...10^{-3}$ of the dose applied due to a bee sting (unit dose). In the following, increasing doses of allergen are injected, obtained by multiplying the initial dose with factors 1, 2, 4, 7, 10, 20, 40, 70, 100, 200, 400, etc. The series of injections ends reaching a value of about twice the normal dose. (For wasp

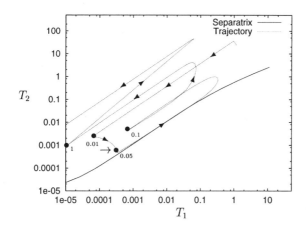

Figure 6: Unsuccessful treatment. The treatment fails since the dose of the second injection of allergen (indicated by an arrow) is too large. The separatrix is not crossed and the response to the last injection is strongly Th2-dominated. ($\mu_1 = 1$, $\mu_2 = 2$, $v = 8$, $c = 0.0001$, $\alpha = 10$, $\phi = 0.95$).

venom the therapy may be continued up to the 10...20 fold of the unit dose.)

After this initial phase the therapy is continued in larger distances of weeks or months. A successful therapy leads to a response which is less allergic or not allergic at all. The improvement is however only temporary, the therapy could be repeated after some years. The doses and the schedule of administration are found empirically, there is no general theoretical guidance.

Our model offers a theoretical explanation of hyposensitization in the frame of nonlinear dynamics. Generally speaking, the state space of system (18-20) separates into regions where the response to an injection of an unit dose of allergen is either Th1- or Th2-dominated. Obviously, for the parameter setting introduced above, the naive state is in the region where the response to a unit dose of allergen is Th2-dominated. The allergen injections make the system non-autonomous. A suitable sequence of injections may drive the system to cross the boundary towards the region of Th1-dominated response, we call this boundary 'dynamical separatrix'.

Figures 3 and 4 show the concentrations of naive cells, Th1- and Th2-cells, and allergen as a function of time for identical parameters but different therapy. Figure 3 shows a successful therapy which leads to a Th1-dominated response – a hyposensitized state. The therapy in Fig. 4 fails since the doses of allergen increase too fast.

Figures 5 and 6 show projections of the trajectories into the Th1-Th2 plane for a successful and an unsuccessful therapy, respectively. In the successful case the dynamical separatrix between the regions of different response is crossed, in the latter case the system stays in the Th2-dominated region.

The strategy of hyposensitization exploits the competition between autocrine and cross-reactive interactions for different ratios of Th1/Th2 concentrations. In our model hyposensitization has three major steps:

(i) In the beginning, T-cell concentrations are small so that both cross-suppression and autocrine effects can be neglected against the stimulation by the (small) cytokine background. The parameter ϕ alone decides which T-cell population is preferably produced by the naive cells ($\phi < 1$ favours Th1). The allergic state is characterized by a choice of μ_1 and μ_2 controlling the cross-suppression such that the response to a small (resp. large) dose of allergen is Th1- (resp. Th2-) dominated.

(ii) Injections of slowly increasing doses of allergen (starting with a small dose provoking a Th1-response) drive the system gradually to higher concentrations of T-cells for which stimulation by autocrine effects is not longer negligible compared with stimulation by background cytokines. However, since Th1-cells are already preferred, the ratio Th1/Th2 can be further increased so that even at concentrations for which cross-suppression (favouring Th2) becomes important, the response to a large dose of allergen is Th1-dominated.

(iii) The hyposensitized state is only temporary; the system relaxes finally to a state which shows an allergic response – unless the therapy is repeated.

The above discussion suggests that a larger cytokine background (leading *a priori* to higher T-cell concentrations) makes hyposensitization more difficult or even impossible. Figure 7 shows a modified schedule for a higher cytokine background.

5   T-cell memory

To include T-cell memory (see e.g. Bradley *et al.* (1993),[4] and Croft (1994)[5]) we introduce Th1- and Th2-memory cells, denoted as M_1 and M_2, which are produced by stimulated naive cells in the presence of cytokines, similar to normal Th1- and Th2-cells. In our model, memory T-cells have a longer life

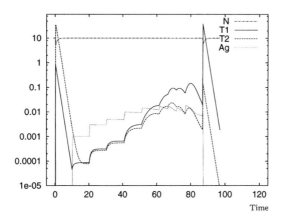

239

Time	0	10	20	30	40	50	60	65	70	75	85
Ag–Dose	1	0.001	0.002	0.002	0.005	0.005	0.005	0.005	0.005	0.01	1

Figure 7: Successful hyposensitization for a higher cytokine background. A higher cytokine background makes it necessary to inject the allergen in very slowly increasing doses at shorter time intervals. ($c = 0.001$, other parameters as in Fig. 5).

time than normal T-cells. Stimulated by allergen they produce Th1- and Th2-cells. We assume for simplicity that –in contrast to naive cells– stimulation of memory cells need not the presence of cytokines. After adiabatic elimination of the cytokine variables and introducing reduced dimensions the equations governing the dynamics read

$$\frac{dN}{dt} = -N + \alpha - N\,Ag\,(\frac{T_1}{1 + \mu_2\,T_2} + c) - \phi\,N\,Ag\,(T_2 + c), \quad (22)$$

$$\frac{dT_1}{dt} = -T_1 + v\,N\,Ag\,(\frac{T_1}{1 + \mu_2\,T_2} + c) + v\,f\,M_1\,Ag, \quad (23)$$

$$\frac{dM_1}{dt} = -\varepsilon M_1 + u\,N\,Ag\,(\frac{T_1}{1 + \mu_2\,T_2} + c) - f\,M_1\,Ag, \quad (24)$$

$$\frac{dT_2}{dt} = -T_2 + v\,\phi\,N\,Ag\,\frac{T_2 + c}{1 + \mu_1\,\dfrac{T_1}{1 + \mu_2\,T_2}} + v\,f\,M_2\,Ag, \quad (25)$$

$$\frac{dM_2}{dt} = -\varepsilon M_2 + u\,\phi\,N\,Ag\,(T_2 + c) - f\,M_2\,Ag, \quad (26)$$

240

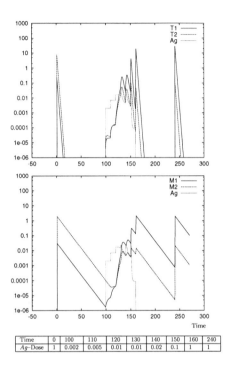

Time	0	100	110	120	130	140	150	160	240
Ag-Dose	1	0.002	0.005	0.01	0.01	0.02	0.1	1	1

Figure 8: Hyposensitization in the model with T-cell memory. The therapy starts 100 time units after the primary response. 80 time units after the therapy the hyposensitized state was still preserved. ($\mu_1 = 1$, $\mu_2 = 3$, $v = 8$, $c = 0.0001$, $\alpha = 10$, $\phi = 0.9$, $u = 0.5$, $f = 10$, $\epsilon = 0.1$).

$$\frac{dAg}{dt} = \xi(t) - Ag\,(T_1 + T_2). \tag{27}$$

Here, $1/\varepsilon$ is the life time of memory cells measured in units of the life time of normal T-cells. We have varied ε between 0.1 and 0.01. u determines how many memory cells are produced by a stimulated naive cell. Since memory T-cells are a smaller population than normal T-cells we chose $u = 0.5$. f determines the efficiency of the stimulation of memory cells by allergen and thus the efficiency of the secondary response.

Including memory cells into our description amounts in two effects, illustrated in Figs. 8 and 9. The persistence of the hyposensitized state after a successful therapy is prolonged in the order of the life time of memory cells, but still hyposensitization is temporary. Furthermore, the doses of allergen applied during therapy should be chosen more carefully. Because of the increased

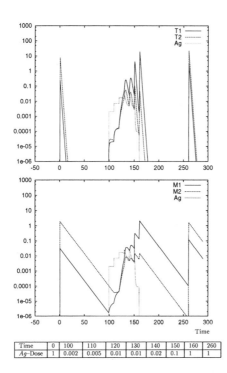

Time	0	100	110	120	130	140	150	160	260
Ag-Dose	1	0.002	0.005	0.01	0.01	0.02	0.1	1	1

Figure 9: Hyposensitization in the model with T-cell memory is still a *transient* phenomenon. After the same therapy as in Fig. 7 we wait 100 time units to inject the allergen dose $\xi = 1$. A Th2-dominated response is observed so that the treatment should be repeated.

secondary response the therapy should start with smaller doses.

6 Concluding Remarks

Modeling Th1-Th2 regulation needs of course simplification of a very complex scheme. Our intention was to explain hyposensitization therapy as a dynamic effect of Th1-Th2 regulation. The dynamic behaviour of the model and experimental results [6,7,8] support this hypothesis. The desensitized state is a temporary state, *in vivo* and in the developed model. Model simulations show, for example, that even one missed allergen injection during therapy or one too high allergen dose may endanger the success of hyposensitization. In case of a missed allergen injection it may be helpful to decrease the dose of the next allergen injection. Other simulations show that the application of cytokines, which meets however difficult problems in practice, seems to be a conceivable way in therapy. The mathematical model offers the possibility to test variations in the schedule of administration, e.g. changing doses and intervals between injections, which may help *in practice* to optimize therapies.

Th1-Th2 regulation, especially the switch from the dominance of one T-cell subset to another, has a special role not only in allergy. Altered profiles of Th1-Th2 type cytokine production may be associated also with other diseases in humans such as candidiasis, AIDS, Leishmania and autoimmune diseases. Antigen administration could provoke a switch in the T-cell dominance in these diseases, too. For a very recent survey of current research the reader is referred to Abbas *et al.* (1997)[9] and to Romagnani and collaborators (1997).[10,11,12]

We are confident that mathematical models of T-cell regulation could provide assistance in understanding these diseases and may help to develop and to optimize therapeutic strategies.

References

1. L.M. Lichtenstein, *Allergy and the immune system*, Scientif. American special: The immune system, 85–93 (Sept. 1993).
2. F. Powrie and R.L. Coffman, *Cytokine regulation of T–cell function: potential for therapeutic intervention*, Immunol. Today **14**, 270–274 (1993).
3. T.R. Mosmann and S. Sad, *The expanding universe of T–cell subsets: Th1, Th2 and more*, Immunol. Today **17**, 138–146 (1996).
4. L. M. Bradley, M. Croft, and S. L. Swain, *T–cell memory: new perspectives*, Immunol. Today **14**, 197–199 (1993).
5. M. Croft, *Activation of naive, memory and effector T cells*, Curr. Opin. Immunol. **6**, 431–437 (1994).
6. H. Secrist, C.J. Chelen, Y. Wen, J.D. Marshall, and D.T. Umetsu, *Allergen immunotherapy decreases Interleukin 4 production in CD4$^+$ T cells from allergic individuals*, J. Exp. Med. **178**, 2123–2130 (1993).
7. S. Romagnani, *Regulation of Th2 development in allergy*, Curr. Opin. Immunol. **6**, 838–846 (1994).
8. V.A. Varney, Q. Hamid, M. Gaga, S. Ying, M. Jacobsen, A.J. Frew, A.B. Kay, and S.R. Durham, *Influence of grass pollen immunotherapy on cellular infiltration and cytokine mRNA expression during allergen-induced late-phase cutaneous responses*, J. Clin. Invest. **92**, 644–651 (1993).
9. A.K. Abbas, K.M. Murphy, and A. Sher, *Functional diversity of helper T lymphocytes*, Nature **383**, 787–793 (1996).
10. S. Romagnani, *The Th1/Th2 paradigm*, Immunol. Today **18**, 263–266 (1997).

11. S. Romagnani, *The Th1/Th2 Paradigm in Disease* (Springer-Verlag, Heidelberg 1997).
12. P. Parronchi, E. Maggi, R. Manetti, M.P. Piccinni, S. Sampognaro, A. Annunziato, L. Giannarini, L. Beloni, and S. Romagnani, *T lymphocytes and subpopulations: Involvement of Th2 cells in allergic diseases* In J. Ring, H. Behrendt, and D. Vieluf, editors, *New Trends in Allergy*, volume IV, pages 131–136 (Springer-Verlag, Berlin Heidelberg 1997).

Articles

PHASE TRANSITIONS IN A PROBABILISTIC CELLULAR AUTOMATON WITH TWO ABSORBING STATES

FRANCO BAGNOLI

Dipartimento di Matematica Applicata "G. Sansone"
Università di Firenze, via S. Marta, 3 I-50139, Firenze, Italy
INFN and INFM sez. di Firenze
e-mail: bagnoli@dma.unifi.it

NINO BOCCARA

DRECAM/SPEC, CE-Saclay, F-91191 Gif-sur-Yvette Cedex, France
Department of Physics, University of Illinois, Chicago, USA
e-mail: nboccara@amoco.saclay.cea.fr and boccara@uic.edu

PAOLO PALMERINI

Dipartimento di Fisica, Università di Firenze
Largo E. Fermi 2, I-50125 Firenze, Italy
e-mail: palmerini@dma.unifi.it

Abstract

We study the phase diagram and the critical behavior of a one-dimensional radius-1 two-state totalistic probabilistic cellular automaton having two absorbing states. This system exhibits a first-order phase transition between the fully occupied state and the empty state, two second-order phase transitions between a partially occupied state and either the fully occupied state or the empty state, and a second-order damage-spreading phase transition. It is found that all the second-order phase transitions have the same critical behavior as the directed percolation model. The mean-field approximation gives a rather good qualitative description of all these phase transitions.

1 Introduction

Probabilistic cellular automata (PCA) have been widely used to model a variety of systems with local interactions in physics, chemistry, biology and social sciences. [1,2,3,4,5] Many of these models exhibit transcritical bifurcations which may be viewed as second-order phase transitions. It has been conjectured [6,7] that all second-order phase transitions from an "active" state characterized by a nonzero scalar order parameter to a nondegenerate "absorbing" state provided that (i) the interactions are short range and translation invariant, (ii) the probability at which the transition takes place is strictly positive, and (iii) there are no multicritical points belong to the universality class of directed

247

percolation (DP). In the Domany–Kinzel (DK) cellular automaton,[8,9] whi
is the simplest PCA of this type, Martins et al.[10] discovered a new "phase
These authors considered two initial random configurations differing only
one site, and studied their evolution with identical realizations of the stocha
tic noise. The site at which the initial configurations differ may be regard
as a "damage" and the question they addressed was whether this damage w
eventually heal or spread. The new phase, they called "chaotic", is characte
ized by a nonzero density of damaged sites. In a recent paper, Grassberger
(see also Bagnoli (1996)[12]) showed that this new phase transition also belo
to the universality class of directed percolation.

More recently, Hinrichsen[13] stressed that in presence of multiple absorbir
states with symmetric weight, the universality class changes from DP to
parity conservation (PC) class, with quite different exponents.

The damage-spreading phase transition can be related to usual chaot
properties (i.e. positivity of maximal Lyapunov exponent) of dynamical sy
tems.[14] The boundary of the damaged phase depend in general on the a
gorithm used for the implementation of the evolution rule. Recently, how
ever, Hinrichsen[15] formulated an objective, algorithm-independent definition
damage spreading transitions. In particular, the algorithm with maximal co
relation between the random numbers used in the updating (i.e. using only on
random number) allows the computation of the minimum boundary of dam
age spreading transitions: inside this region a damage will spread whicheve
algorithm is used.

In this paper, we study the phase diagram and the critical behavior c
a one-dimensional radius-1 totalistic PCA having two absorbing states, an
which exhibits a multicritical point. In the DK model, which is a two-inpu
PCA characterized by the transition probabilities $P(1|00) = 1$, $P(1|01) =$
$P(1|10) = p_1$ and $P(1|11) = p_2$, the system has two absorbing states for $p_2 = 1$
and exhibit a first-order phase transition at $p_1 = 1/2$. We may therefore
expect that our system will have a nontrivial phase diagram. Moreover, th
discretization of differential equations lead naturally to radius-1 rules (or three
inputs rules) and we feel that it would be useful to have a better knowledge o
the phenomenology of simple models formulated in terms of radius-1 PCA.

2   The model

The evolution rule of our model is defined as follows. Let $s(i,t)$ denotes the
state of the i-th cell at time t, and $\theta(i,t) = s(i-1,t) + s(i,t) + s(i+1,t)$ the

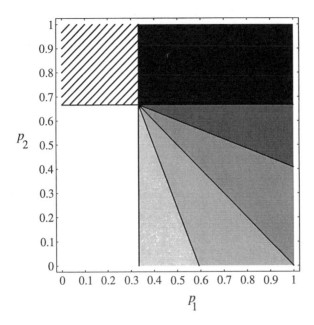

Figure 1: Mean-field phase diagram for the density c of active sites for $p_3 = 1$, as described by Equation (1). The gray levels indicates the asymptotic value of c; contours for $c = 0$, 1/4, 1/2, 3/4 and 1. The dashed region marks coexistence of phases. The first order phase boundary depends on the initial condition.

sum of the cells in the neighborhood, then

$$s(i, t + 1) = \begin{cases} 0, & \text{if } \theta(i, t) = 0, \\ X_1, & \text{if } \theta(i, t) = 1, \\ X_2, & \text{if } \theta(i, t) = 2, \\ X_3, & \text{if } \theta(i, t) = 3, \end{cases}$$

where X_j ($j = 1, 2, 3$) is a Bernoulli random variable equal to 1 with probability p_j, and to 0 with probability $1 - p_j$. That is, the transition probabilities of this model are $P(0|000) = 1$, $P(1|001) = P(1|010) = P(1|100) = p_1$, $P(1|011) = P(1|101) = P(1|110) = p_2$, and $P(1|111) = p_3$. Cell i is said to be "empty" at time t if $s(i, t) = 0$, and "occupied" if $s(i, t) = 1$. The state in which all cells are empty is a fixed point (absorbing state) of the dynamics. If $p_3 = 1$, the state in which all cells are occupied is also a fixed point. In this work we study mainly the case $p_3 = 1$, delaying the study of the case $p_3 < 1$ to section 5.

If there exists a stable "active" state such that the asymptotic density c

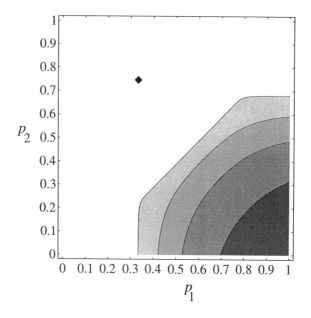

Figure 2: Mean-field damage-spreading phase diagram for $p_3 = 1$. The phase diagram has been obtained numerically iterating Equation (2). Levels of damage are $h = 0, 1/8, 1/4, 3/8$ and $1/3$. The isolated dot marks the multicritical point.

of occupied cells is neither zero nor one, then the model will exhibit various bifurcations between these different states as the parameters p_1 and p_2 vary. This totalistic PCA can be viewed as a simple model of opinion formation. It assumes that our own opinion and the opinion of our nearest neighbors have equal weights. The role of social pressure is twofold. If there is homogeneity of opinions, then one does not change his mind (absorbing states), otherwise one can agree or disagree with the majority with a certain probability.

Let us assume that one starts with a random configuration half filled with 0's and 1's. On the line $p_1 + p_2 = 1$, if p_1 is small, to first order in p_1, clusters of ones perform symmetric random walks, and have equal probabilities to shrink or to grow. Slightly off this line, the random walks are no more symmetric, and according to whether p_1 is greater or less than $1 - p_2$, all the cells will eventually be either occupied or empty. $p_1 + p_2 = 1$ is, therefore, a first-order transition line. This line, however, cannot extend to $p_1 = 1$ since, for $p_2 = 0$, our model is similar to the diluted XOR rule (Rule 90) of the DK model. We thus expect that our model will exhibit, for a certain critical value of p_1,

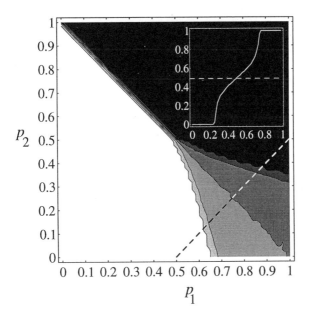

Figure 3: Phase diagram for the density of active sites c obtained by direct simulations. One run was performed. Lattice size: 10000, number of time steps: 10000, resolution: 64 different values for both p_1 and p_2 ($p_3 = 1$), color codes as in Figure 1. The initial density is $c_0 = 0.5$. The inset represents the density profile along the dashed line. Two critical phase transitions are evident.

a second-order phase transition coinciding with the damage-spreading phase transition. For $p_1 = 1$ and $p_2 = 0$, frustration is large, and our model is just the modulo-2 rule (Rule 150) as in the DK model.

For $p_3 = 1$, the model is symmetric under the exchange $p_1 \leftrightarrow 1 - p_2$, and $0 \leftrightarrow 1$. Thus, along the line $p_2 = 1 - p_1$ the two absorbing states 0 and 1 have equal weight, and, according to Hinrichsen,[13] this transition should belong to the PC universality class.

3   Mean-field approximation

In order to have a qualitative idea of its behavior, we first study our model within the mean-field approximation. If $c(t)$ denotes the density of occupied cells at time t, we have

$$c(t+1) = \quad 3p_1 c(t) \left(1 - c(t)\right)^2 + \\ 3p_2 c^2(t) \left(1 - c(t)\right) + c^3(t). \tag{1}$$

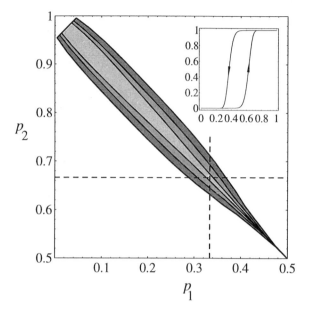

Figure 4: Profile of the hysteresis region for several values of the noise ε and relaxation time T. Data obtained from the local structure approximation of order $l = 6$; the regions corresponds to the intersections of the hysteresis cycle with $c = 0.5$ (horizontal dashed line in the inset). The hysteresis boundary lines smoothly join at $p_1 = 0$, $p_2 = 1$ (not represented). Darker to lighter areas correspond to $T = 500$, $\varepsilon = 0.0001$; $T = 1000$, $\varepsilon = 0.0001$; $T = 500$, $\varepsilon = 0.001$. The dashed lines represent the mean-field hysteresis region. The inset represents the cycle along a line parallel to the diagonal $p_1 = p_2$.

This one-dimensional map has three fixed points:

$$0, \quad 1, \quad \text{and} \quad c^* = \frac{3p_1 - 1}{1 + 3p_1 - 3p_2}.$$

0 is stable if $p_1 < \frac{1}{3}$, 1 is stable if $p_2 > \frac{2}{3}$, and c^* is stable if $p_1 > \frac{1}{3}$ and $p_2 < \frac{2}{3}$. In the (p_1, p_2)-parameter space, the bifurcations along the lines

$$\begin{aligned} p_1 &= \tfrac{1}{3}, & 0 \leq p_2 &\leq \tfrac{2}{3}, \\ \tfrac{1}{3} \leq p_1 &\leq 1, & p_2 &= \tfrac{2}{3}, \end{aligned}$$

may be viewed as second-order transition lines; the first one between the active and empty states, and the second one between the active and fully occupied states.

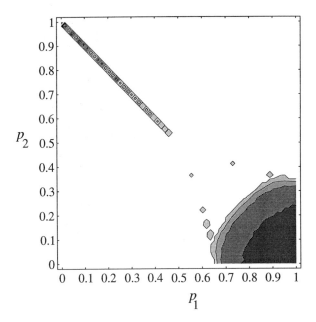

Figure 5: Phase diagram for the damage spreading from direct numerical simulations. Numerical values as in Figure 3, color codes as in Figure 2. Traces of the second order phase transitions are present; they join to the first order ($c_0 = 0.5$) phase boundary.

In the domain defined by

$$0 \le p_1 < \frac{1}{3} \quad \text{and} \quad \frac{2}{3} < p_2 \le 1$$

fixed points 0 and 1 are both stable. Their basins of attraction are, respectively, the semi-open intervals $[0, c^*[$ and $]c^*, 1]$. Therefore, if we start from a uniformly distributed random value of c, as time t goes to infinity, $c(t)$ tends to 0 with probability c^*, and to 1 with probability $1 - c^*$. Since, for $p_1 + p_2 = 1$, $c^* = \frac{1}{2}$, the line defined by

$$p_1 + p_2 = 1, \quad \text{with} \quad 0 \le p_1 < \frac{1}{3} \quad \text{and} \quad \frac{2}{3} < p_2 \le 1$$

is similar to a first-order transition line, although we cannot define a free energy for our dynamical system.

In general, first order phase transitions are associated to the presence of an hysteresis cycle, due to the coexistence of two phases, both in equilibrium.

254

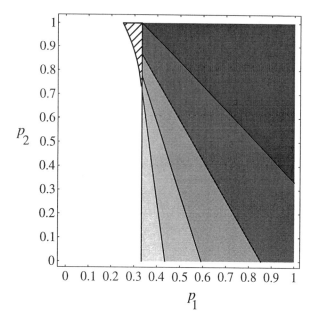

Figure 6: Mean-field phase diagram for $p_3 = 0$. Color codes as in Figure 1.

The problem of a proper definition of such a cycle is rather subtle: since this model presents absorbing states, it is out of equilibrium and the ergodicity is always broken, even for finite-size systems. Thus, once the system settles into one of the two absorbing states, it never exits, even if it is not stable. One can circumvent this problem adding a little of noise, i.e. setting $P(0|000) = 1 - P(1|111) = \varepsilon$. This assumption, however, brings the model into the class of equilibrium models, thus forbidding the presence of a true phase transition in one dimension. However, if ε is small, the system can get trapped for a long time T in a metastable region near an absorbing state. The time T is also a function of the system size L. Thus, we have to perform carefully the limits $L \to \infty$, $T \to \infty$ and $\varepsilon \to 0$. In practice, we have observed that the intensity of noise ε strongly affects the amplitude of the hysteresis region, while, for a system size L sufficiently large, there exists a large interval of possible time length T. The results from numerical simulations are reported in section 4.

In the mean-field approximation the hysteresis region ranges from $0 < p_1 < 1/3$, $p_2 = 2/3$ to $p_1 = 1/3$, $2/3 < p_2 < 1$. The mean-field phase diagram for the density is shown in Figure 1.

One can write down the mean-field equation for the damage spreading aking into consideration all possible local configurations of the two lattices. Let us denote by $\eta(t)$ the density of damaged sites at time t. The evolution equation for η depends on correlations among sites, i.e. on the order of the mean field for the density. Using the simplest factorization for the density, $\eta(t)$ depends on $c(t)$. The evolution equation for the minimum damage η, i.e. the damage when the evolution of the two replicas is computed using only one random number, is given by

$$\eta(t+1) = \sum_{\substack{s_1 s_2 s_3 \\ h_1 h_2 h_3}} \pi(c(t), s_1 s_2 s_3) \pi(\eta(t), h_1 h_2 h_3) \cdot \left| P(1|s_1 s_2 s_3) - P(0|s_1 s_2 s_3 \oplus h_1 h_2 h_3) \right| \tag{2}$$

where

$$\pi(\alpha, x_1 x_2 x_3) = \alpha^{x_1 + x_2 + x_3} (1 - \alpha)^{3 - x_1 + x_2 + x_3}$$

and the symbol $\oplus$ represents the bitwise sum modulus two (XOR) of two Boolean configurations. The value for $c(t)$ is given by mapping (1).

We have numerically iterated Equation (2). The resulting mean-field phase diagram for the damage is shown in Figure 2.

4   Numerical simulations

We used the fragment method [16] to determine the phase diagram. Figure 3 represent a plot for the asymptotic density c of occupied cells when the initial fraction of the occupied cells is equal to 0.5. The scenario is qualitatively the same as predicted by the mean-field analysis. In the vicinity of the point $(p_1, p_2) = (0, 1)$ we observe a discontinuous transition from $c = 0$ to $c = 1$, while close to $(p_1, p_2) = (1, 0)$ the transition is continuous. The two second-order phase-transition lines from the non fully occupied states to either the fully occupied state or the empty state are symmetric and the critical behavior of the respective order parameters, $1 - c$ and c, are the same. We have checked that the critical exponent β for these transitions far from the crossing point are numerically the same of DP.

These two transition lines meet on the diagonal $p_1 + p_2 = 1$ and become a first-order transition line. Crossing the phase boundaries on a line parallel to the diagonal $p_1 = p_2$, the density c exhibits two critical transitions, as shown in the inset of Figure 3. Approaching the crossing point, the critical regions of the two transitions vanishes, the transition itself become sharper and the corrections to scaling increase.

At the crossing point, around $(0.5, 0.5)$, the two attractors have symmetrical weight. If we relabel couples of 1's with the symbol 1, couples of 0's with

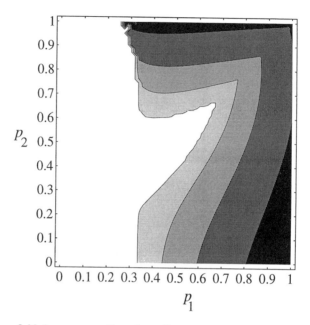

Figure 7: Mean-field damage-spreading phase diagram, for $p_3 = 0$. The diagram has been obtained numerically iterating Equation (2).

the symbol 0 and couples 01 or 01 with the symbol A, we fulfill the condition stated by Hinrichsen[13] to have symmetric absorbing states, whose cluster are always separated by an active (A) layer. We have performed preliminary measurements of the exponent β for the density of A couples along the line $p_1 + p_2 = 1$, and obtained $\beta \simeq 0.65(5)$, which is consistent with the values for the PC process reported in Grassberger et al. (1984),[17] Grassberger (1989).[18]. The transition point is at $p_1 = 0.460(2)$, which defines the position of the tricritical point.

In order to study the presence of an hysteresis region, we performed several scanning of the first-order transition using the local structure approximation.[19] We cut the phase diagram with a line parallel to the diagonal $p_1 = p_2$, increasing the value of p_1 and p_2 after a given relaxation time up to $p_2 = 1$; then reverting the scanning up to $p_1 = 0$. The results are reported in Figure 4. The size of the hysteresis region grows with the level of the noise ε. Preliminary simulations show that the relaxation time T and the noise level ε scales as $\varepsilon T^{1.434} = $ const.

The phase diagram for the damage spreading is shown in Figure 5. Outside

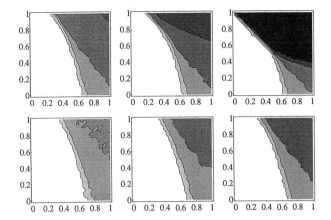

Figure 8: Cut of the density phase space along the p_1-p_2 plane for different values of p_3. From left to right and for bottom to top: $p_3 = 0.0$, 0.2, 0.4, 0.6, 0.8 and 1.0. Color codes as in Figure 1

the true damage-spreading region, there appear small damaged domains on the other phase boundaries. This is due either to the divergence of the relaxation time (second-order transitions) or to the fact that a small difference in the initial configuration can drive the system to a different absorbing state (first-order transitions).

The "chaotic" domain near the point $(p_1, p_2) = (1, 0)$ is stable regardless of the initial density. Note that on the line $p_2 = 0$ its boundary coincides with the transition line from the active state to the empty state. Using an argument similar to the one in Bagnoli (1986).[12] would prove that the derivatives of the two boundary curves coincide, that is, the chaotic phase exhibits a *reentrant* behavior. Here again, the symmetry of the phase diagram implies a similar behavior on the line $p_1 = 1$.

5 Extended phase diagram

It is interesting to study our model when the transition probability $P(1|111) = p_3 < 1$. In this case, there is only one absorbing state.

5.1 Mean-field approximation

For some values of p_3, the mean-field phase diagram still exhibits a first-order phase transition. Let us first examine in detail the particular case $p_3 = 0$.

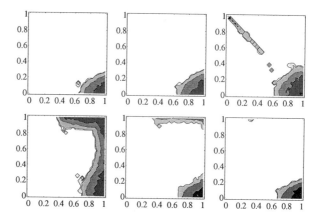

Figure 9: Cut of the damage-spreading phase space along the p_1-p_2 plane for several values of p_3. From left to right and for bottom to top: $p_3 = 0.0$, 0.2, 0.4, 0.6, 0.8 and 1.0. Color codes as in Figure 2.

As for $p_3 = 1$, the solution $c = 0$ looses its stability at $p = 1/3$. The other solutions for the density c are the roots of the quadratic equation

$$c^2(p_1 - p_2) + c(p_2 - 2p_1) + p_1 - \frac{1}{3} = 0.$$

These two roots are real if $q^2 + 4(p_1 - p_2)/3 \geq 0$. This is always the case if $p_1 > 1/3$, while for $p_1 < 1/3$, p_2 must lie outside the interval $\left[\frac{2}{3}(1 - \sqrt{1 - 3p_1}), \frac{2}{3}(1 + \sqrt{1 - 3p_1})\right]$. Since the density c has to be such that $0 \leq c \leq 1$, only the condition $1 \geq p_2 > \frac{2}{3}(1 + \sqrt{1 - 3p_1})$ is meaningful. These solutions are stable. Therefore, there is a domain in which two stable solutions coexist. The boundaries of this domain are the three straight lines $p_1 = \frac{1}{3}$, $p_2 = 1$ and $p_2 = \frac{2}{3}(1 + \sqrt{1 - 3p_1})$, where for this last line p_2 belongs to the interval $\left[\frac{1}{4}, \frac{1}{3}\right]$. When, inside the domain, we approach the point $(p_1, p_2) = \left(\frac{1}{3}, \frac{2}{3}\right)$, the density c tends continuously to zero, which shows that, at this point, the transition is second-order. The corresponding phase diagram is represented in Figure 6.

For the chaotic phase, the boundary of the domain of stability is shown in Figure 7. Close to the point $(p_1, p_2) = (1, 0)$ the boundary is almost the same as for $p_3 = 1$ but there exists another domain close to the point $(p_1, p_2) = (1, 1)$ which did not exist for $p_3 = 1$. In the vicinity of $(1, 1)$, very small clusters, which tend to grow, are then fragmented because $p_3 = 0$. We have also studied the case $0 < p_3 < 1$.

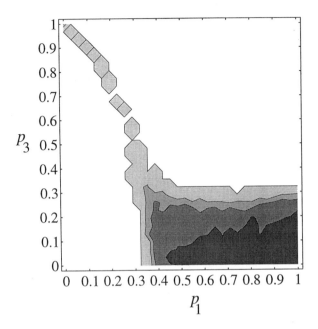

Figure 10: Cut of the density phase space along the p_1-p_3 plane for $p_2 = 1$. Color codes as in Figure 1.

5.2 Numerical simulations

Our numerical simulations give results in good qualitative agreement with the mean-field approximation. Figure 8 shows different phase diagrams in the (p_1, p_2)-plane corresponding to different values of p_3. As expected, the boundary of the active phase $(c \neq 0)$ for $p_2 = 0$ does not vary much with p_3 since, in this case, the probability to find a cluster of three or more one is vanishing. No first order phase transition for the density is apparent by these simulations.

Figure 9 shows the domain of stability of the chaotic phase in the (p_1, p_2)-plane for different values of p_3. The domain of stability of the chaotic phase in the (p_1, p_3)-plane for $p_2 = 1$ is represented in Figure 10. One can see that there are damage-spreading domains corresponding to the density phase boundaries.

6 Conclusion

We have studied a radius-1 totalistic PCA whose transition probabilities are given by $P(1|000) = 0$, $P(1|001) = P(1|010) = P(1|100) = p_1$, $P(1|011) = P(1|101) = P(1|110) = p_2$, and $P(1|111) = p_3$, within the framework of the mean-field approximation and using numerical simulations.

For $p_3 = 1$, the system has two absorbing states and four different phases: an empty state in which no cell is occupied, a fully occupied state, an active phase in which only a fraction of the cells are occupied, and a chaotic phase. This system exhibits:

1. A first-order phase transition from the fully occupied state to the empty state.

2. Two second-order phase transitions from the active state to the fully occupied state and from the active state to the empty state.

3. A second-order phase transition from the active state to the chaotic state.

For $p_3 < 1$, our numerical simulations show that there is no first-order phase transition while the mean-field approximation still predicts the existence of such a transition.

References

1. D. Farmer, T. Toffoli, and S. Wolfram, editors, *Cellular Automata*, Los Alamos Interdisciplinary Workshop, (North Holland, Amsterdam 1984).
2. S. Wolfram, editor, *Theory and Applications of Cellular Automata*, (World Scientific, Singapore 1987).
3. P. Manneville, N. Boccara, G. Vichniac, and R. Bidaux, editors, *Cellular Automata and Modeling of Complex Physical Systems*, Les Houches Workshop, (Springer, Heidelberg 1989).
4. H. Gutowitz, editor, *Cellular Automata: Theory and Experiments*, Los Alamos Workshop, (North-Holland, Amsterdam 1990).
5. N. Boccara, E. Goles, S. Martínez, and P. Picco, editors, *Cellular Automata and Cooperative Phenomena*, Les Houches Workshop, (Kluwer, Berlin 1993).
6. H.K. Jensen, Z. Phys. B **42**, 152 (1981).
7. P. Grassberger, Z. Phys. B **47**, 365 (1982).
8. E. Kinzel and W. Domany, Phys. Rev. Lett. **53** (1984).
9. W. Kinzel, Z. Phys. B **58** (1985).
10. M.L. Martins, H.F.V. de Resende, C. Tsallis, and A.C.N. de Magalhaes, Phys. Rev. Lett. **66**, 2045 (1991).

11. P. Grassberger, J. Stat. Phys. **79**, 13 (1995).
12. F. Bagnoli, J. Stat. Phys. **85**, 151 (1996).
13. H. Hinrichsen, Phys. Rev. E **55**, 219 (1997), http://xxx.lanl.gov/cond-mat/9608065.
14. F. Bagnoli, R. Rectman, and S. Ruffo, Physics Letters A **172**, 34 (1992).
15. H. Hinrichsen, J.S. Weitz, and E. Domany, J. Phys. E , (in press) (1997).
16. F. Bagnoli, P. Palmerini, and R. Rechtman, Phys. Rev. E **55**, 3970 (1997).
17. P. Grassberger, F. Krause and T. von der Twer, J. Phys. A: Math. Gen. **17**, L105 (1984).
18. P. Grassberger, J. Phys. A: Math. Gen. **22**, L1103 (1989).
19. H.A. Gutowitz, J.D. Viktor, and B.W. Knight, Physica **28**, 18 (1987).

HIGH COOPERATIVITY AS ORIGIN OF PATTERN COMPLEXITY

ROBIN ENGELHARDT
Center for Chaos And Turbulence Studies
Chem.Lab. III, Dept. of Chemistry
University of Copenhagen, Universitetsparken 5
Dk - 2100 Copenhagen Ø, Denmark
email: kel3re@unidhp.uni-c.dk

Abstract

Biological pattern formation is a process which so far has gone largely unexplained. Experimental biologists favour mechanisms in which morphogenetic gradients are the source of positional information, but among theoreticians the pattern forming processes are believed to some extend to be dependent upon true symmetry breaking processes, which can occur in most nonlinear control systems, of which the Turing mechanism is an example. It is argued here that the common fundamental feature of gene control systems is a high degree of nonlinearity and thus cooperativity, and that this mechanism also is important for the pattern forming processes in early evolution.

1 Pattern Formation in Turing Systems

The process of biological pattern formation is to a high extend depending on nonlinear dynamics in the complex biological control systems. [2] High nonlinear on-off control in the defining kinetics of the system, is a central feature of "active gene control" and also a prerequsite for the interpretation of positional information. For a particular pattern formation mechanism, Turing's mechanism, [3] it is shown that the increment of cooperativity in the defining reaction rates - expressed by the Hill number - greatly facilitates pattern formation. The degree of nonlinear control in such systems is usually meassured by the Hill number in the kinetic equations. Usually, effective Hill numbers in such studies have been taken to be less that approx. 3, but in genetic control systems it is common to have substantially larger Hill numbers, often apparently in excess of 8. This indicates that high cooperativity makes gene systems prone to effectiv Turing type pattern formation, as well as time-oscillations and cue control. Thus, the study of nonlinear dynamics and the properties of nonlinear control systems, are equally important for the understanding of the emergence of patterns in multicellular systems, as are studies of protein structure or the geometry of protein-gene interactions.

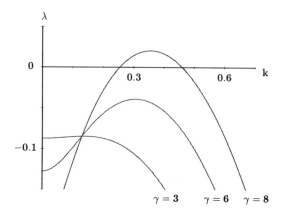

Figure 1: The figure shows the standard dispersion plot for Turing systems, showing the real part of the eigenvalue as a function of the spatial mode κ. When increasing the Hill constant γ (but keeping the diffusion constant), as is shown for three different cases, the eigenvalues become positive, and Turing patterns are formed. Thus, large cooperativity (large Hill constants) results in in much easier pattern formation.

Note in figure 1 that the normal dispersion equation for these kind of reaction-diffusion equations yields an increasing Turing region with increasing Hill number. The eigenvalue λ is calculated as a function of wave number κ as usual, but with the extension of plotting a set of such curves for increasing values of the Hill constant γ, and with a constant ratio of the diffusion coefficients, $D_-/D_+ = 5$. This shows that pattern formation by Turings Mechanism is facilitated by increasing cooperativity and thus high effective Hill constants γ in the defining rates. Such an increase in nonlinearity has occured during evolution in gene control systems, for example where the promoter of the gene has developed several binding sites for gene controlling proteins. Cooperativity with effective γ in excess of 8 has been recorded experimentally for several different gene control systems. Such high Hill constants may have developed under evolutionary pressure, as they are necessary for accurate control in other contexts, even in single cells, but such control systems also become increasingly prone to create spontaneous pattern formation by Turing's mechanism. In all the models investigated, we obtain an expression of the form

$$\frac{D_-}{D_+} > \frac{\sqrt{z}+1}{\sqrt{z}-1} \tag{1}$$

Table 1: Effect on the critical diffusion coefficient when increasing the Hill constant.

z	D_-/D_+
2	5.83
4	3.00
8	2.09
16	1.67

where z is function of γ, usually a linear function at least for suitable parameter values, and thus z increases with increasing cooperativity γ. It is tempting to suggest, as a conjecture, that this will always obtain, but even with occasional exceptions to this rule the above results show that a substantial class of models become Turing systems with increasing cooperativity. Thus the usually stated requirement of having effective diffusion coefficients differ by almost an order of magnitude, which is often found in systems with small Hill constants, is relaxed in gene control systems where much larger effective Hill constants have been recorded.

Table 1 shows that the ratio of diffusion coefficients D_-/D_+ in a Turing system approaches one when effective Hill constant increases according to Eq. (1). In this equation z is a function of the effective Hill constant γ and for a number of mechanisms it has been shown here that $z \simeq \gamma$ at least for suitable parameter combinations. We conclude that Turing pattern formation may be much more feasible in actual biological systems, if the mechanism is connected to the gene control system with high Hill numbers, than would be expected from studies of, for example, inorganic model systems, or experimental realizations of such inorganic systems, as the latter systems rarely have Hill numbers exceeding 2. This discovery thus supplements earlier discussions of the feasibility of a substantial ratio between effective diffusion constants in Turing's mechanism. [1,5]

2   Early Biological Morphogenesis

It is believed that the homebox class of genes is somehow connected to the origin of pattern formation in multicellular systems. [4] However, the basis of pattern formation emerged well before the Cambrian explosion some 600 million years ago. Many "defining" characteristics for the metazoa are widely distributed among unicellular eukaryotes. Positional information seems to be present already in the ciliates, and fungi and plants arose much earlier than the

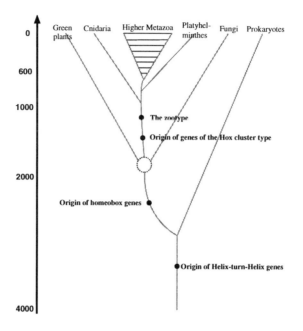

Figure 2: A highly schematical figure of the tree of evolution with special reference to the origin of pattern formation.

metazoans, with their own system of pattern forming processes. It is stressed here that the actual pattern forming processes observed in early evolution have a more common feature than a single "pattern protogene" and thus the homeobox genes are not "pattern" genes as such, but rather convenient control genes linked to another set of fundamental pattern forming processes. It is suggested that the key feature of these genes is their ability to create highly cooperative, on-off dynamics. Such dynamics is well known to be capable of generating a rich variety of well controlable patterns.

Figure 2 shows the evolutionary tree with special reference to the origin of pattern formation (the time scale for the splits and gene origins should be seen only as a rough guideline). One role the homebox genes - or their predecessors, the genes for helix-turn-helix proteins - may have played in this context, is to provide a system with very high cooperativity. Initially in evolution such a system may be used simply to read subtle differences of concentrations, quite possible within a single cell. Indeed, several processes need fine tuned control in which the system is required to respond in a nearly on-off manner when concentrations are varied with less than a factor of 2. The main point then is that a gene system which is capable of discriminating small differences in concentration is prone to generate cue control, chemical oscillations and Turing type patterns as well.

The connection between the homeobox genes and the fundamental pattern forming processes in evolution may thus **not be that of an ancestral pattern gene** which has developed into a class of such genes. Rather, the homeobox genes have the ability to create proteins which bind to the promoter region of another gene, and thereby be capable to be involved in highly cooperative control. Thus we will stress again, in summary, that the study of nonlinear control systems is equally important for the understanding of the emergence of patterns in multicellular systems, as are studies of protein structure or the geometry of protein-gene interactions.

References

1. A. Hunding and P.G. Sørensen, *Size adaptation of turing prepatterns*, J.Math.Biol. **26**, 27–39 (1988).
2. A. Hunding and R. Engelhardt, *Early biological morphogenesis and nonlinear dynamics*, J. Theor. Biol. **173** 401–413 (1995).
3. A.M. Turing, *On the chemical basis of morphogenesis*, Phil.Trans.Roy.Soc., Ser. B **237** 37–72 (1952).
4. L.G. Harrison, *What is the status of reaction-diffusion theory thirty-four years after turing?*, J. Theor. Biol. **125**, 369–384 (1987).
5. G.R.S. Métens et al., *Pattern formation in bistable chemical systems*, Proceedings of the IMACS third International Conference (Copenhagen, August 1994).

STRUCTURED MATHEMATICAL MODELS FOR DYNAMICS OF MICROBIAL GROWTH: FIRST ORDER AND DISCRETE TIME DELAYS

PETER GÖTZ

Fachgebiet Bioverfahrenstechnik, Technische Universität Berlin,
Ackerstrasse 71-76, 13355 Berlin

Abstract

Kinetic models for microbial growth are usually derived from pseudo steady-state assumptions. Dynamic conditions in the environment of the microorganisms may lead to complex responses within the metabolism of the organisms. Observing the macroscopic variables like biomass or substrate concentration, the effect of changes of metabolism within the cells results in transient behavior, which can be summarized under the term "biological inertia".

This work shows two different modeling approaches for this effect, introducing a growth governing intracellular concentration, which changes after a time delay according to environmental changes. These delays are modeled by a first order and a discrete time delay respectively. Parameter estimation fom experimental data and model discrimination shows, that the discrete time delay is more suited for data representation. This is related to the concept of "maturation" of intracellular compounds, between induction of intracellular processes and the resulting activity often a discrete time delay is observed.

1 Introduction

Technical applications of processes using microorganisms require detailed knowledge about microbial growth and biodegradation of substrates. For a better understanding of the process dynamics, mathematical models for biodegradation and microbial growth are valuable tools for improving bioreactor design, predictions of operating conditions and process control. Usually biochemical engineers apply unstructured models derived from mass balances, describing the biomass and substrate concentration in the reactor.

Corresponding model parameters can be estimated from experiments in batch, fed-batch or continuous cultures. Evaluating these parameters for the same microbial system from these different reactor systems commonly leads to discrepancies. [1] This is due to different process dynamics causing various intracellular events which are not included in the model. A powerful tool for identifying such effects is a phase plane plot of the experimental data (Biomass vs. Substrate). The existence of intersections within the trajectories of the two variables shows the demand for additional variables to describe the process since these intersections arise from a projection of a higher dimensional phase space.[2] Another sign for the model being incomplete are hysteresis effects in a plot of the specific growth rate vs. the substrate concentration.[3,4] These time-dependent deviations from single valued functions like Monod kinetics are also caused by dynamic intracellular events.

This leads to the evolution of structured models incorporating these events in the model. Some tools for handling dynamic phenomena are introduction of empirical activity functions,[5] of balanceable quantities like enzyme[6] or RNA concentrations[7,6] and of transfer functions.[8,9] An alternative way to modelling metabolic regulation is known as cybernetic approach.[10]

The incorporation of time effects in an engineering description of biological non-steady-state processes is commonly accomplished by a first order delay for the specific growth rate or a growth rate determining variable.[7,6] Models for time effects being set up by researchers with a biological background often use a discrete delay.[11,12,13] This stems from the fact that after their synthesis is initiated, complex biological structures need a certain time for maturation before they become active.

In this work experimentally observed dynamic phenomena during phenol degradation will be presented and structured mathematical models including balances for intracellular quantities and time delays will be developed.

The microorganisms *Pseudomonas putida* DSM 549 and *Trichosporon beigelii* (self isolate) were used in the experiments, both organisms are able to use phenol as sole energy and carbon source. Growth rates of the microorganisms in steady states of a two-stage continuous culture with phenol as energy and carbon source can be described by a Haldane type of equation (a star denotes quantities originating from steady states)

$$\mu^\star = \mu_{max} \frac{S}{K_S + S + \dfrac{S^2}{K_I}}. \tag{1}$$

From mass balances for substrate and biomass follows the system of differential equations describing transient states of a continuous culture

$$\frac{\mathrm{d}X}{\mathrm{d}t} = (\mu - D)X,$$
$$\frac{\mathrm{d}S}{\mathrm{d}t} = (S_0 - S)D - \frac{\mu X}{Y'_{X/S}} - m_S X. \tag{2}$$

Included into the model are a theoretical maximal yield coefficient $Y'_{X/S}$ and a maintenance coefficient m_S. Solving Eqns. (2) for the transient state

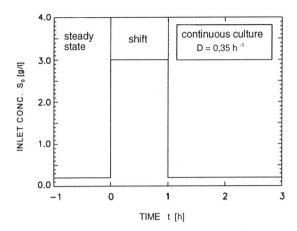

Figure 1: Time course of phenol inlet concentration during shift-experiment.

of imposing an one hour shift of the inlet phenol concentration on the reactor (Fig. 1) operating in steady state, predicts a washout for the microbial population. The measured results for seven identical shift experiments are shown in Fig. 2. The most important observation is that there is no wash-out, instead the population returns after about 8 hours to the steady-state it had before the shift.

The observed deviation from predicted reactor behavior indicates, that the model is incomplete. A common approach for expanding a model for microbial growth is the addition of structure by including the activity of intracellular growth limiting components, for example ribosomes. The fact, that experimental verification of the model prediction for the shift experiment using steady state parameters failed for both organisms, *Pseudomonas putida* and *Trichosporon beigelii*, can be judged as being a clue for the cause of this phenomenon. As there seems to be a common source of this effect in procaryotic and eucaryotic organisms, both using different pathways for phenol metabolism, the ribosome, which is common to both organisms, could be the cause of the dynamic effect.

Ribosomes play a central role in protein biosynthesis, their content within the cell is growth rate dependent. Ribosomes consist of about 60% RNA and ribosomal RNA accounts for the largest amount of RNA in the cell, therefore the measurement of RNA is used as a measure for the amount of ribosomes in the cell.

272

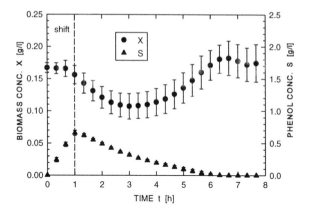

Figure 2: Data from 7 shift-experiments (*P.putida*, upper curve: Biomass, lower curve: Substrate).

Including the RNA-concentration into the model must not change the properties of the operating points measured in steady states of the continuous culture. Therefore new growth kinetics are defined for the expanded model

$$\mu = kR \frac{S}{K_M + S + \dfrac{S^2}{K_N}}. \tag{3}$$

For compatibility between the steady-state and the dynamic model, we need

$$\mu|_{R=R^\star} = \mu^\star. \tag{4}$$

Setting Eq. (1) equal to Eq. (3) and rewriting for the steady state RNA concentration $R^\star$ yields

$$R^\star = \frac{\mu_{\max}}{k} \frac{K_M + S + \dfrac{S^2}{K_N}}{K_S + S + \dfrac{S^2}{K_I}} \tag{5}$$

This type of equation allows the representation of experimental data on steady state RNA-concentrations in the cell. Another observation from dynamic experiments is the influence of phenol concentration on the measured biomass yield. In the model this can be accounted for by a maintenance coefficient m_S which depends on the substrate concentration, hence the expanded set of model equations is

$$\frac{dX}{dt} = (\mu - D)X$$

$$\frac{dS}{dt} = (S_0 - S)D - \frac{\mu X}{Y'_{X/S}} - (m_{S0} + m_{S1}S)X \tag{6}$$

$$\frac{dR}{dt} = r_R - \mu R$$

The mathematical formulation of the rate of RNA synthesis r_R will be the topic of the following considerations.

The rate of change of RNA content during transient conditions can be thought of being governed by a setpoint control.[7] This setpoint is taken from the relationship Eq. (5).

To account for dynamic effects within the formation rate of RNA r_R in Eq. (6) this rate is introduced as a sum of the steady state formation rate $r_R^\star$ and the rate due to dynamic effects r_R^{dyn}. Since excess ribosomes are not degraded in the cell, no negative formation rate is observed

$$r_R = r_R^\star + r_R^{\text{dyn}}. \tag{7}$$

The steady state formation rate of RNA, $r_R^\star$, is found by introducing Eq. (7) into the differential equation for R (Eq. (6)) with the time differential and the dynamic component rRdyn being equal to zero in steady state. Rearranging yields

$$r_R^\star = \mu R. \tag{8}$$

The dynamic component of the RNA formation rate is assumed to be a function of the difference between the amount of RNA which could be used under steady state conditions and the actual RNA content.[7]

2   First order delay

A simple and widely used way of formulating the dynamic component r_R^{dyn} is a first order delay. For the process of adaptation, a relaxation time constant τ is introduced[7]

$$r_R^{\text{dyn}} = \frac{1}{\tau}(R^\star - R). \tag{9}$$

The data for biomass concentration from the shift experiment are used to estimate the model parameters by minimizing a quality function. The parameters, which are already known from theroretical considerations ($Y_{X/S}$) and steady state experiments ($\mu_{\max}$, K_S, K_I, m_{S0}) are fixed. Only parameters related to dynamic processes (τ, k, K_M, K_N and m_{S1}) are optimized by a simplex method. With these parameters, the model achieves a good fit of data with a mean relative error of 3.2%. Although this result seems reasonable a further interpretation shows two problems:

- Mathematical simulation of the RNA content during the shift experiment results in values of about 80% RNA with respect to total cell mass, which is in contradiction to observations.

- The identified value for the time constant τ is 13.1 h. This does not fit the framework of relaxation times due to changes of RNA concentrations which are in the order of 1 h.

Because of these contradictions, the concept of a first order delay for modeling the observed dynamic effects has to be abandoned.

3 First order and discrete delays

Closer inspection of the synthesis of ribosomes reveals, that between the production of the building blocks of a ribosome and the first activity of the generated ribosome a maturation time is observed. This delay is caused by the assembly of ribosomal proteins and rRNA. A first order delay may therefore not be suited to describe the process of ribosome synthesis properly. An alternative modeling approach is the inclusion of this time of maturation by using a discrete delay. This approach uses two additional state variables, the inactive building blocks of ribosomal RNA N and the active ribosomal RNA R in mature ribosomes. The time delay between production of N and activity of R is introduced as delay time Θ, the tilde ($\tilde{\ }$) denotes variables subject to the discrete time delay $(t - \Theta)$. Using a first order delay (cf. Eq. (9)) for the dynamics of production of building blocks N and eliminating the variable N yields

$$r_R^{\mathrm{dyn}} = \left(\tilde{\mu}\tilde{R}e^{\tilde{\mu}\Theta} + \tfrac{1}{\tau}(\tilde{R}^\star - \tilde{R}) \right) e^{-\int_{t-\Theta}^{t} \mu(t)\mathrm{d}t}.$$

$$\quad\quad 1 \quad\quad\quad\quad 2 \quad\quad\quad\quad 3 \quad\quad\quad\quad\quad 4$$

(10)

The expressions in Eq. (10) are interpreted as follows:

1. Dynamic component of production rate of active ribosomes / ribosomal RNA;

2. Steady state component of building block / nucleotide RNA production rate at time $(t - \Theta)$;

Table 1: Parameters for time delay model Eqs. (6), (7) and (10)

$\mu_{\max}$	$0.54\,\mathrm{h}^{-1}$
K_S	$1.5\,10^{-3}\,\mathrm{gl}^{-1}$
K_I	$0.18\,\mathrm{gl}^{-1}$
$Y'_{X/S}$	$0.94\,\mathrm{gg}^{-1}$
m_S	$0.054\,\mathrm{g(gh)}^{-1}$
m_{S1}	$1.98\,\mathrm{l(gh)}^{-1}$
k	$4.62\,\mathrm{g(gh)}^{-1}$
K_M	$7.81\,10^{-4}\,\mathrm{gl}^{-1}$
K_N	$0.299\,\mathrm{gl}^{-1}$
τ	$1.45\,10^{-2}\,\mathrm{h} = 52\mathrm{s}$
Θ	$0.517\,\mathrm{h}$
mean relative error	5.1%

3. Dynamic component of nucleotide RNA production rate, a function of the deviation of ribosomal RNA from the setpoint at time $(t - \Theta)$

4. Dilution of nucleotide RNA content caused by growth during the maturation time interval between $(t - \Theta)$ and t.

The fit of the model with parameters in Table 1 to the data is shown in Fig. 3. Since the agreement is good, the parameter values will be further examined. The value for τ seems reasonable since nucleotides are small molecules and can be synthesized easily. The value of m_{S1} agrees well with results from different evaluations. The magnitudes of the parameters K_M and K_N agree with the postulates from compatibility of dynamic and steady states and together with the parameter k they can be used to predict the steady state RNA content of the cells (cf. Eq. (6)).

Model discrimination resulted in the first order/discrete delay model being suited for fitting the experimental data. Interpretation of the parameter values identified for the model agrees with the mechanistic framework of this modeling approach. The number of model parameters is small considering the number of constraints from independent measurements, which are satisfied by the model. This reinforces the hypothesis of ribosomes being the cause for the observed biological inertia effects.

The problems in this type of model are the rates of change in the variables, especially the RNA content of cells (Fig. 4). Although the order of magnitude of the simulated results seems reasonable, the oscillations in the model are

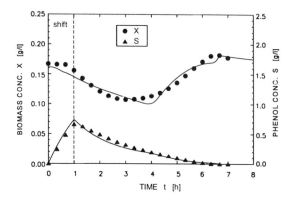

Figure 3: Shift-experiment: Comparison between discrete delay model results and experimental data.

too steep and the experimentally observed dynamics are much smaller. The reason lies within the discrete delay time in the model. In reality this delay time for maturation of ribosomes will not be represented by a single number but by a distribution of maturation times. There are different ways to represent such a distribution, one possibility is the introduction of a chain of linear differential equations,[8] another is the use of a empirical transfer function.[9] The implementation and evaluation of such modeling approaches will be the topic of further work.

References

1. P. Götz, *Einsatz strukturierter Modelle zur Beschreibung dynamischer Zustnde mikrobieller Populationen unter Bercksichtigung der biologischen Trägheit* TU Berlin, Fachbereich Verfahrenstechnik und Energietechnik, Dissertation (1992).

2. J.A. Howell and M.G. Jones, *The development of a two-component operon-type theory model for phenol degradation* AIChE Symposium Series **209**, 122-128 (1981).

3. P. Fischer, *Reaktionskinetische Untersuchungen zum mikrobiellen Phenolabbau* TU Berlin, Fachbereich Lebensmittel- und Biotechnologie, Dissertation (1989).

4. R.D. Tanner, *Departures from the monod fermentation growth (kinetic) model*, Proceedings of the Seventh International Biotechnology Symposium, (New Dehli, India 1984).

277

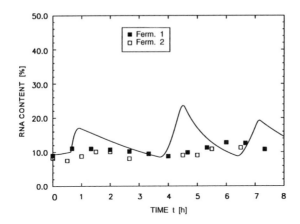

Figure 4: RNA content during shift-experiment: Comparison between discrete delay model results and data from two experiments.

5. C.T. Chi and J.A. Howell, *Transient behavior of a continuous stirred tank biological reactor utilizing phenol as an inhibitory substrate*, Biotech. and Bioeng. **18**, 63-80 (1976).

6. J.M. Romanovsky,N.V. Stepanova and D.S. Chernavsky, *Kinetische Modelle in der Biophysik* (VEB Gustav Fischer Verlag, Jena 1974).

7. J.A. Roels, *Energetics and kinetics in biotechnology* (Elsevier Biomedical Press, Amsterdam 1983).

8. N. MacDonald, *Time lags in biological models. Lecture Notes in Biomathematics*, (Springer, Berlin 1978).

9. N.M. Wang and G. Stephanopoulos, *A new approach to bioprocess identification and modeling*, Biotech. and Bioeng. Symp. **14**, 635-656 (1984).

10. D. Ramkrishna, J.V. Straight, A. Narang and A.E. Konopka, *Modeling of microbial processes. The status of the cybernetic approach* In *Harnessing Biotechnology for the 21st Century*, M.R. Ladisch, A. Bose (eds.), (ACS, Washington DC 1992).

11. U. an der Heiden, *Delays in physiological systems* J. Math. Biology **8** 345-364 (1979).

12. M.C. Mackey and L. Glass, *Oscillation and chaos in physiological control systems*, Science **19**, 287-289 (1977).

13. A. Shibata and N. Sait, *Time delays and chaos in two competing species*, Mathematical Biosciences **51**, 199-211 (1980).

STABILITY OF MICROBIAL MIXED CULTURE BIOPROCESSES

MARKUS RARBACH

Fachgebiet Bioverfahrenstechnik, Technische Universität Berlin
Ackerstr. 71-76, 13355 Berlin

PETER GÖTZ

Fachgebiet Bioverfahrenstechnik, Technische Universität Berlin
Ackerstr. 71-76, 13355 Berlin

Abstract

In nature the coexistence of different microorganisms in mixed cultures is the usual source of microbial activity. If phenomena in mixed cultures may be explained by studying pure cultures is the aim of this work. Choosing a model system consisting of two microbial species with commensalistic and mutualistic interactions, first the pure cultures are studied experimentally and modelled mathematically. Synthesis of the two models for pure cultures yields a model for the mixed culture. Stability analysis of this model exhibits the existence of multiple steady states and bifurcations. Experimental model verification in a continuous mixed culture shows good model prediction of steady states, but the model fails for dynamic states. Including a time delay related to the interaction between the two microorganisms which accounts for the "biological inertia" of adaptation to changes in the environment leads to an improved model prediction for dynamic processes.

Technical use of combined metabolic capacity in microbial mixed cultures is limited to environmental engineering and food engineering. In continuous waste water treatment processes it is essential to use the multitude of reactions catalyzed by different microorganisms. This is especially true for industrial waste water treatment, where problems arise from inhibitory or toxic substrates. Under these circumstances a process is sensitive to disturbances. Wash - out of one microbial species from the network of different organisms can lead to a breakdown of efficiency.

In food engineering many traditional production processes use mixed cultures:

- Milk processing (e.g. cheese or fermented beverages from milk)

- Alcohol production (e.g. sour mash Whisky)

- Bread production (sourdough).

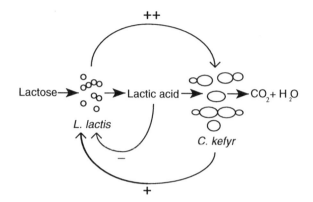

Figure 1: Interactions in a mixed culture of *L. lactis* and *C.kefyr*. Symbols: ++: essential positive influence; +: positive influence; −: negative influence.

All these processes have their roots in naturally occuring stable equilibria of coexistence for different microorganisms.

In biotechnological production mixed cultures are used only in research due to problems in process stability, reproducability, process control, downstream processing and constant product quality. As an example the production of the antibiotic Biphenomycin A in mixed culture leads to a 60-fold product yield compared to a single organism culture.[1] A technical realisation of this process is hindered by the aforementioned problems.

Basic research on reaction engineering analysis of biochemical reaction networks and mixed cultures reveals phenomena, which can help understanding experimental results.[2,3,4] Mathematical modelling of such systems using experimental model verification has been under investigation only in a few cases.[5] In the following there will an example of analyzing a mixed culture system using the tools of reaction engineering.

In basic research we used a mixed culture of *Lactococcus lactis* and *Candida kefyr* with substrate Lactose. Interactions in this system are shown in Fig. 1. *C. kefyr* is not able to metabolize Lactose, therefore the metabolic activity of *L. lactis* is essential for growth of the yeast *C. kefyr*. The activity of *C. kefyr* has a positive effect on *L. lactis*, because the inhibitory metabolic product (Lactic acid) from the bacterium is used up by the yeast. In order to describe the system mathematically, the various reaction kinetics of the organisms are measured separately in batch and fed-batch cultivations. The parameters from these investigations will then be used to model the continuous mixed culture. As an example Fig. 2 shows a fed-batch fermentation with *L.lactis*, which is

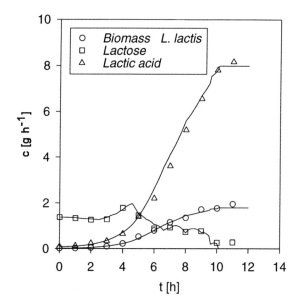

Figure 2: Fed-batch cultivation of *L. lactis*: Comparison of model simulation and experimental data.

designed to reveal the reaction kinetics of product inhibition. By adjusting the feed rate (Fig. 3), an approximately constant high level of lactose in the bioreactor is maintained, whereas the product concentration is allowed to change over a broad range of concentrations. The growth of biomass is slowed down at high product concentrations, confirming the assumption of inhibition and yield reduction by the product lactic acid. Model structure is derived from such experiments and model parameters are estimated using experimental data, for the example *L. lactis* we use the following assumptions (cf. Fig. 4 and Eq. (1)):

- Lactose S from the reactor medium is transported into the *L. lactis* cells by an inducible transport system (Permease E) with rate $r_{tr,L}$ and thereby converted to intracellular lactose S_i. Transport is inhibited by lactic acid (product) concentration P and regulated by intracellular lactose concentration S_i.

- S_i is used with rate $r_{S_i,L}$ for growth of *L. lactis* (μ_L) and for maintenance metabolism of cells (m).

- Growth μ_L for *L. lactis* follows Monod-kinetics using intracellular sub-

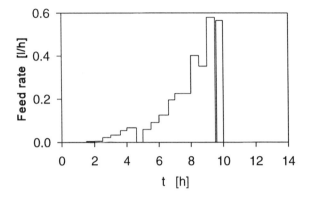

Figure 3: Feed rate for fed-batch cultivation of $L.$ $lactis.$

strate S_i.

- The transport system E is produced from biomass L, the production rate r_E has a constitutive part $r_{E,0}$ which is always active, and an inducible part which is induced by S_i and regulated by a feedback repression using E.

$$r_E = r_{E,0} + r_{E,1} \frac{S_i}{K_{E,ind} + S_i \left(\dfrac{E}{K_{rep}} \right)},$$

$$r_{tr,L} = r_{tr,L,\max} \frac{S}{K_M + S \left(1 + \dfrac{P}{K_P} \right)} \frac{K_I}{K_I + S_i},$$

$$\mu_L = \mu_{L,\max} \frac{S_i}{K_S + S_i},$$

$$r_{S_i,L} = \frac{\mu_L}{Y_{X/S}} + (m_0 + m_1 P) \frac{S_i}{K_S + S_i},$$

(1)

where

L concentration of $L.$ $lactis$, g/l;

S concentration of lactose, g/l;

E intracellular concentration of transport enzyme, g/g;

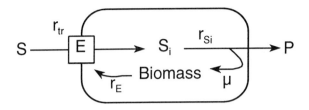

Figure 4: Model structure for *L. lactis* metabolism.

S_i intracellular concentration of substrate lactose, g/g;

P metabolic product (lactic acid) concentration, g/l.

From mass balances we get the following differential equations for the continuous culture (Dilution rate D, inlet lactose concentration S_0) with respect to *L. lactis*

$$\frac{\mathrm{d}L}{\mathrm{d}t} = (\mu_L - D)L,$$

$$\frac{\mathrm{d}S}{\mathrm{d}t} = (S_0 - S)D - r_{tr,L}EL,$$

$$\frac{\mathrm{d}E}{\mathrm{d}t} = r_E - k_d E - \mu_L E,$$

$$\frac{\mathrm{d}S_i}{\mathrm{d}t} = r_{tr,L}E - r_{S_i,L} - \mu_L S_i.$$

(2)

In a mixed culture, the product lactic acid used by the yeast *C. kefyr* (C) is

$$\frac{\mathrm{d}P}{\mathrm{d}t} = Y_{P/S}r_{S_i,L}L - DP - r_{tr,C}C.$$

(3)

The mass balances for *C. kefyr* are therefore coupled to the mass balances of *L. lactis* by the metabolic product lactic acid, which is transported into the cells of *C. kefyr* (conversion to P_i) and then used for growth (μ_C)

$$\frac{\mathrm{d}C}{\mathrm{d}t} = (\mu_C - D)X,$$

$$\frac{\mathrm{d}P_i}{\mathrm{d}t} = r_{tr,L} - r_{P_i,L} - \mu_C P_i.$$

(4)

Model parameters from the aforementioned independent experiments in batch and fed-batch cultures allow the numerical solution of the system of equations. Model predictions are generated for stationary and dynamical states

284

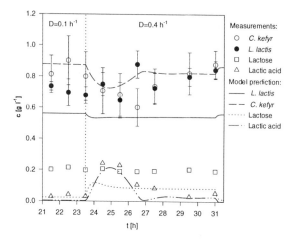

Figure 5: Shift in dilution rate D: Comparison of model prediction and measurements.

of the continuous mixed culture. Experimental verification of these predictions (Fig. 5) leads to the final model. From these dynamic experiments a time delay in *C. kefyr* metabolism was observed, which required an expansion of the model. For this expansion the maximal transport rate $P \rightarrow P_i$ for *C. kefyr* is modified. In order to keep the singular points for the observable variables unchanged, a first order time delay with respect to the maximal transport capacity $r_{tr,\max,C}$ is used (additional information on time delays in biological systems can be found elsewhere in this book). The final model includes one more differential equation in addition to Eqns. (2–4), replacing the parameter $r_{tr,\max,C}$ with the variable ν and introducing a setpoint ν_{set}

$$\frac{\mathrm{d}\nu}{\mathrm{d}t} = \frac{1}{\tau}(\nu_{\text{set}} - \nu) \qquad \nu_{\text{set}} = \nu_{\max}\frac{P}{K_\nu + P}. \qquad (5)$$

The mathematical model can be analysed for multiplicity and stability of operating points. Operating points are a subset of the steady states, which are the singularities of the system of differential equations. Singular points with no physical meaning (e.g. negative concentrations) or trivial solutions ($C = 0$, $L = 0$) are not considered. Setting up the Jacobian of the system and evaluating the eigenvalues at the operating points yields information on stability of these points. Using the dilution rate D as control parameter, a plot of stable and unstable stationary operating points shows multiplicity of

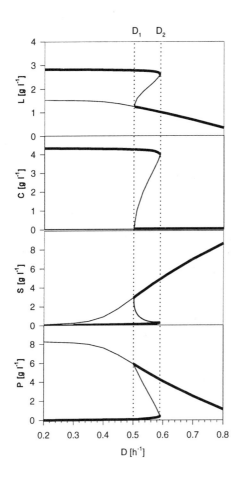

Figure 6: Control parameter dilution rate D: Multiple steady states in continuous mixed culture. Variables: L, Biomass conc. *L. lactis*; C, Biomass conc. *C. kefyr*; S, Conc. Lactose; P Conc. Lactic acid.

steady states in the range between D_1 and D_2 (Fig. 6). Thick lines are stable operating points, thin lines denote unstable states. The behavior of this system can be compared to the ignition and extinction phenomenon in non-isothermal chemical reactors. [6] At the dilution rates D_1 and D_2 the number of steady states changes, this phenomenon is the well known bifurcation. Interpretation of Fig. 6 with respect to the process shows, that hysteresis effects will occur in this interval. If the dilution rate is increased above D_2, the yeast will wash out. In order to prevent a wash out of the yeast, a decrease of dilution rate below D_2 will not neccesarily bring the system back to the previous operating point, because between D_1 and D_2 two stable steady states exist. Only a decrease of dilution rate below D_1 will stabilize the mixed culture.

This example shows that mathematical modeling and model analysis can result in valuable information on complex biological processes. Operating and troubleshooting of technical applications is simplified, especially in continuous processes which employ many different microbial species (wastewater treatment). Expanding technical applications of mixed cultures to production of biotechnological high-value products makes a thorough process analysis indispensable. For constant product quality, dynamic effects (changes in populations, induction of metabolic pathways etc.) must be predicted from a model and controlled properly. Introducing time delays to account for such effects has interesting consequences for the mathematical model. Although equations like Eq. (5) will not change the values of the observable variables at a singular point, the stability properties and the basin of attraction will change. Introducing higher order time delays, from numerical experiments the occurence of a Hopf-bifurcation was observed, a transition of a stable stationary state to a stable dynamic state (limit cycle). Further work will be directed to developing the mathematical framework for process analysis and experimental verification of stability properties of operating points.

References

1. M. Ezaki, M. Iwami, M. Yamashita, T. Komori, K. Umehara and H. Imanaka, *Biphenomycin A Production by a Mixed Culture*, Appl. And Envir. Microbiol. **58**, 3879-3882 (1992).
2. N.S. Panikov, *Mechanistic mathematical models of microbial growth in bioreactors and in natural soils: Explanation of complex phenomena*, Mathematics and Computers in Simulation **42**, 179-186 (1996).
3. M. Sheintuch, *Dynamics of commensalistic systems with self- and cross-inhibition*, Biotech. Bioeng. **22**, 2557-2577 (1980).
4. J.W. Stucki, *Stability analysis of biochemical systems - a practical guide*,

Prog. Biophys. Molec. Biol. **33**, 99-187 (1978).

5. N. Noisommit-Rizzi, *Abbau von Sulfanilsaeure mit einer definierten Mischpopulation: Experimentelle Beobachtungen und Entwurf eines kinetischen Modells*, Dissertation (TU Berlin 1994).

6. J.D. Chung and G. Stephanopoulos, *On physiological multiplicity and population heterogeneity of biological systems*, Chemical Engineering Science **51**, 1509-1521 (1996).

SIMULATING THE IMMUNE RESPONSE ON A DISTRIBUTED PARALLEL COMPUTER

F. CASTIGLIONE

Dep. of Mathematics, Univ. of Catania

E-mail: castif@dipmat.unict.it

M. BERNASCHI

IBM DSSC, Roma

E-mail: massimo@vnet.ibm.com

S. SUCCI

IAC CNR, Roma

E-mail: succi@sparc1.iac.rm.cnr.it

Abstract

The application of ideas and methods of statistical mechanics to problems of biological relevance is one of the most promising frontiers of theoretical and computational mathematical physics. [1,2] Among others, the computer simulation of the immune system dynamics stands out as one of the prominent candidates for this type of investigations.

In the recent years immunological research has been drawing increasing benefits from the resort to advanced mathematical modelling on modern computers. [3,4]

Among others, *Cellular Automata* (CA), i.e. fully discrete dynamical systems evolving according to boolean laws, appear to be extremely well suited to computer simulation of biological systems. [5]

A prominent example of immunological CA is represented by the Celada-Seiden automaton, that has proven capable of providing several new insights into the dynamics of the immune system response.

To date, the Celada-Seiden automaton was not in a position to exploit the impressive advances of computer technology, and notably parallel processing, simply because no parallel version of this automaton had been developed yet.

In the present paper we close this gap and describe a parallel version of the Celada-Seiden cellular automaton aimed at simulating the dynamic response of the immune system.

Details on the parallel implementation as well as performance data on the IBM SP2 parallel platform are presented and commented on.

Keywords: Immune Response, Cellular Automata (CA), Parallel Virtual Machine (PVM)

1 Introduction

Most attempts to simulate the behaviour of biological systems are faced with the need of manipulating and processing vast amounts of data. This stems from the "outrageous" number of degrees of freedom available to Nature to accomplish its functional tasks in the biological realm.

The simulation of the dynamic response of the immune system (IS) - one of the most complex systems altogether in Nature - is a particularly challenging task in this respect.

Traditionally, the immune system response is modelled via complex networks of non-linear ordinary differential equations (ODE's) much in the spirit of population dynamics.[6] This approach is certainly valuable, but suffers from a number of drawbacks, primarily a severe exposure to numerical drifts and/or instabilities intrinsic to the floating-point representation of real numbers in digital computers. These instabilities may become particularly offending whenever the number of equations and/or the degree of nonlinearity are increased.

Recently, immunological research (as well as other scientific disciplines) has drawn significant benefits from the resort to Cellular Automata (CA) models, i.e. fully discrete (boolean) dynamical systems, whose fictitious dynamics can be tuned so as to "mimic" the behaviour of the real system at least on a macroscopic scale.[7]

The basic advantage of CA over ODE's is intrinsic numerical stability, which is strictly related the boolean character of the dynamic variables.[5] Many CA models have appeared in the literature which unveiled intriguing analogies between the dynamics of the immune system and the statistical mechanics of lattice spin systems.[8,9,10,11,12] In particular, these simulations show that the notion of phase-transition can be fruitfully exported to the immunological context to describe health/illness state transitions. One of the most prominent attempts to cope with the quest for biological fidelity is the IMMSIM automaton, recently developed by Celada and Seiden.[13,14] IMMSIM still belongs to the class of Immunological Cellular Automata (ICA), but its degree of sophistication sets it apart from simpler ICA in the Ising-like class.[9,11,15]

Needless to say, for all its sophistication, IMMSIM can't keep up with the complexity of real-life immune system. Thus, the inclusion of increasingly complex features (more "actors" and more interaction mechanisms) is always in bad demand. This is the stage where the resort to the impressive advances of computer technology, and notably parallel processing, can be particularly valuable to immunolgical research.

Armed with these considerations, we recently set out to develop a parallel version of the IMMSIM automaton (PARIMM).

First, mandatory, step of this project was to convert the original APL2 code into a high-level language liable to parallel compilation; a number of technical reasons suggested the use of the C language.

Besides, PARIMM was designed according to criteria of *openeness,*and *modularity*, so as to permit smooth upgrades and addition of new features (cells, molecules, interactions, and so on) for future investigations. All the set of simulations presented in this paper are based on the so called *Standard Parameter Set*. These are the reference parameters such that the system proves capable to exhibit the *primary and secondary immune response.*[13,14]

This paper is organized as follows: In section 2 we shall give some brief explanation of the main functionalities of IMMSIM. Section 3 deals with the optimization on the memory management to reduce the memory requirement and in Section 4 we discuss details of the two phase-version of the parallel code (ParImm version 1.0 and 2.0). Finally, in section 4.3, some data pertaining to parallel performance and results are presented and commented on.

2 From biology to computer modeling

The IS is a very complicated system composed by different classes of cells all cooperating to a single goal, the defense of the host organism from attacks of potentially offending invaders.

IMMSIM2.1 [16] is best described at the cell level. The entities involved in the simulation are free to move in a CA-like body rectangle, and all the interactions among them take place on-site, locally, each independently of the others.

The dynamics of the system is mainly governed by these local interactions. Nothing changes in the system (apart from normal fluctuations in the population dynamics) until an antigen (which identifies a generic substance, molecule, virus, bacteria and the like, potentially dangerous for the host organism) gets inside the body (infection). When this happens the system starts the recognition process and the response terminates when the antigen is eliminated by either *humoral* (antibodies) and/or cellular response. During this phase, *memory* cells are produced by clonal division of lymphocites; the memory of what happened sets the system in such a way as to respond promptly to future infections by the *same* antigen.

An infection is simulated by spreading randomly the antigen over the body rectangle as to mimic the injection of antigen in a *mice*.

Among the four kind of cells represented (B,T,AP,PLB), the real chief characters are the B cells (lymphocyte B). Upon stimulation, these produce a clone of PLB (Plasma B cells) which will subsequently secrete antibodies in charge of neutralizing the antigens.

Another peculiar class of cells is the T cell (lymphocyte T Helper); these cells are called helper because they produce the *signal* triggering B cells reproduction via clonal division (i.e. they stimulate B cells). After then both B and T cells will clone to *memory* cells with a longer *lifetime*.

The last kind of cell considered is the APC (Antigen Processing Cell) or *macrophage* representing the non specificity of the system, that is the ability to capture foreign entities without any recognition process.

The function of APC is to catch the antigen, process it (*endocytosis*; the antigen is broken into pieces called peptides and bound, now with a recognition process, the molecule of the *Histocompatibility Complex*, MHC) and presents the MHC/peptide groove on its surface to stimulate T cells. Once this occurs, the T is in a state where it can stimulate B cells to divide.

The same internal processing on captured antigen is also performed by the B cells, but here there is a discriminating-by-affinity recognition when the antigen is captured, meaning that B cells bind antigen epitopes with high specificity by means of its receptors.

In IMMSIM both cell receptors and molecules are binary bit strings of length ℓ. The *binding* between two strings occurs with a certain probability which is function of the *match* (Hamming distance in $\{0, 1\}^{\ell}$) of the two strings.

In other words, when both cells and antigens come close together, they may interact on one condition: their receptors must have a high match (over a certain threshold called *MinMatch*, i.e. minimum match to allow the binding). If this is the case the *probability* to bind is very close to unity.

Every binding event induces a cell-state-change to represent the various phases of the immune response (see Fig 1).

Because we wish to show *specific* response against specific antigen (i.e. specific shape of the epitopes and thus bit string) we expect the cells with a high match receptor to exhibit a higher growth rate.

When B cells clone begin to rise, newborn PLB cells and antibodies (Ab) are produced. The Ab have then the same specificity as the B cell receptor, where PLB come from, so that they are able to bind the antigen with exacly the same strength. In a few time steps of simulation time (several hours or days in real time) the antigen is defeated. After an infection the system shows memory. In fact, subsequent introduction of the same antigen triggers a faster reaction. This is due to the clonal growth of specific memory B and T cells which establish a greater specific defence as compared to the first infection.

The model also implements the *Thymic Selection* of lymphocytes T Helper. This process gives the system the ability to distinguish between *self and non self* molecules and plays a pivotal role in the study of *autoimmune reactions*.

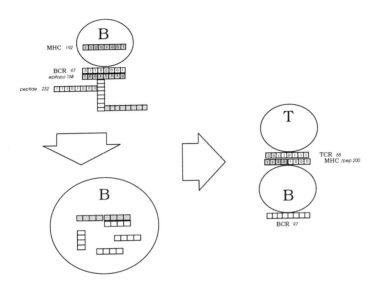

Figure 1: B-Antigen interaction, endocytosis of B cell and B-T interaction

2.1 Model description

In this section we define the phases of the immune system response triggered by events and external conditions.

We start with the first injection of antigens at time step t_0. This may take place anytime after the simulation starts; the system is built to maintain a *steady state* on the global population of cells if no infection is applied. Moreover, initially there no plasma cells or antibodies, nor antibody/antigen complexes in the body. The various steps of the dynamic evolution are:

1. t_0: Injection of antigens (the host has been infected);

2. $t_1 = t_0 + \delta_1$: B Cells bind Ag's with a certain affinity; AP Cells bind Ag's aspecifically;

3. $t_2 = t_1 + \delta_2$: B Cells and AP Cells process internally (MHC bind Ag-peptides) the Ag; $\Rightarrow$ they expose the MHC/peptide groove on the surface;

4. $t_3 = t_2 + \delta_3$: T Cells bind AP and/or B Cells who expose the MHC-complex; $\Rightarrow$ both B and T Cells get stimulated;

Table 1: A subset of the Standard Parameter Set

Standard Parameter Set

Parameter	Value	Meaning
ℓ	8	bit string length
D	16×15	body size
MinMatch	7	minimum match to bind
PtoBind	$(1, 0.05, 0)$	pr to bind (8,7,0-6) match
MHCs	$(137,167)$	MHC molecules
pep/MHC bind	$(1,.5,.25,.125,0)$	pr to bind (4,3,2,1,0) match
$P_{(APC,Ag)}$	0.001	aspecific bind
Ag	$(51,222,17)$	1 epitope + 2 peptides
time to inj Ag	$(0,120)$	injection time
# Ag	500	# Ag injected
#(B=T=APC)	1000	# initial cells

5. $t_4 = t_3 + \delta_4$: Stimulated T Cells divide in T Virgin and T Memory; stimulated B Cells divide in B Memory and PLB Cells;

6. $t_5 = t_4 + \delta_5$: PLB's secrete Ab's;

7. $t_6 = t_5 + \delta_6$: Ab's bind Ag's;

Note that the humoral responce is induced either by two signals; one direct ($Ag \to Bcells$), the other indirect ($Tcells \to Bcells$).

A further contact with the same antigen causes the aforementioned process to be repeated again.

The following plot (Fig 2 up to 4) refers to the population dynamics of B, T and PLB cells, and Ab, Ag and IC molecules in a typical, *Standard*, execution with the parameter as in Table 1.

The parameter set is composed by about ten independent elements, thus giving the model the connotation of a real experimental tool.

In Fig 2 we report the best ten *string-populations* according to receptor strings (for B, PLB and T cells), according to epitope for Ag and receptor for Ab produced by PLB cells. The binary strings are expressed in decimal notation in the figure-titles. The immuno complexes (IC) represent antigens bound by Ab and thus neutralized. The APC are not plotted because they not stimulated to divide and therefore the corresponding population remains in a statististical steady state due to occasional birth and death fluctuations from the initial value.

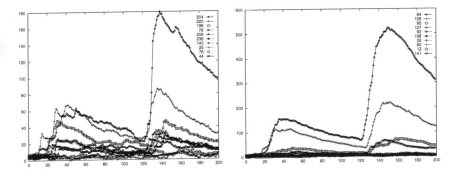

Figure 2: B and T cells

With these parameters settled, the model exhibits the primary and secondary immune response as shown in figures 2-3-4.
The Antigens are injected at time 0 and 120 (Fig 4). After some time (10 time steps) the clonal growth of B cells and also T cells begins to occur (Fig 2).

In Fig 2 the best growth is for the string-population of B cells 204_{10} because of the affinity with the Ag-peptide $51_{10} = 00110011_2$ ($204_{10} = 11001100_2$ is the only perfect match string of 51_{10}). The other string-populations are most likely in the 7-bit match string set (205, 206, 200, 196, 220, 236, 140 and 76). Their growth is strictly depending on the T cells growth. During the response new T and B memory and virgin cells are produced. In Fig 3 one can see the rise of PLB cells coming from stimulated B cells.
The best population of T cells depends on the MHC molecules. In fact they get stimulated if their receptor bind the MHC/peptide groove on the B or AP cells. String 94_{10} (01011110_2) is the perfect match string of 10100001_2 which is composed by half MHC molecule $167_{10} = 10100111_2$ and half Ag-peptide $17_{10} = 00010001_2$ (ref. to Seiden and Celada (1992) [13] for details).
After the antigens disappear (about time step 30), the number of cells decreases. As to the memory cells their reduction is slower because of the longer lifetime.

Note that (Fig 3) the Ab's produced by PLB's, which in turn come from the duplication of B cells, have the same specificity as the originating B cell receptors. Fig 4 shows Ag's and IC's respectively.
It is worth to note that the time required to eliminate the Ag after the

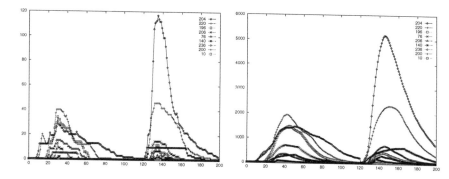

Figure 3: PLB Cells and Antibodies

second injection is about three times shorter.

3 The structure of ParImm

After this brief introduction to the biological environment we start to describe
the details of the ParImm code.

3.1 Dynamic memory allocation

One of the problems in building a complex biological model is the amount
of memory required to store the information which describe the attributes
of the entities. It could be possible to classify the diversity into a pretty
small number of bit-match-classes (clustering), but this would remove useful
information associated with the bit position.

In the early stage of the translation from the original APL2 code into a C
code we adopted a *static* data structure for the entities of the simulation. We
used a number of multidimensional arrays, as many as the classes of entities
represented in the code. The data structures were not homogeneous because
of the different number of attributes of each entity. The attempt to store all
the combination of attributes in a single multidimensional array by lumping
together all indistinguishable cells, had good and bad points.

By lumping together the cells, it is possible, to have in site X, say one-
million-five-hundred-seventeen B cells with receptor (00110011_2) all in "virgin"
state not exposing Ag-peptide. The cost of such information is independent of
the number of cells and requires just four bytes (i.e. the size of an integer on

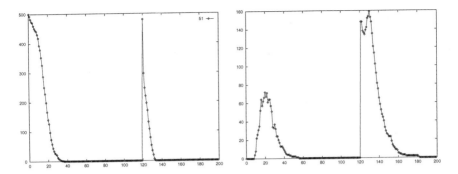

Figure 4: Antigens and Immuno Complexes

most machines). On the other hand we cannot assert *a priori* which information we need at run time because of the stochastic nature of the system. As a
consequence four bytes are required for each possibility. Let's take for example
the B class of cells; we specify the following attributes:

- *cell position within the body* $\in \{0, D-1\}$

- *receptor* $\in \{0, 2^\ell - 1\}$

- *state* $\in \{0, NStates\}$

- *duplication step* $\in \{0, DupSteps\}$

- *MHC/peptide groove* $\in \{0, 2^\ell - 1\}$

this means $2^\ell \times D \times NStates \times DupStep \times 2^\ell \times sizeof(int)$ bytes.

With this scheme for a standard 8-bit system, with fixed *small* antigen and
just one MHC molecule, we need ($\ell = 8, D = 16 \times 15, Nstates = 5, DupStep = 4$), 4.68 MB which is a reasonable amount of memory. However for a 12-bit
system with $D = 30 \times 30, Nstates = 5, DupStep = 4$, and variable (but still
small) number of Ag-epitopes/peptides and MHC molecules, the requirement
increases to 1.17 GB.

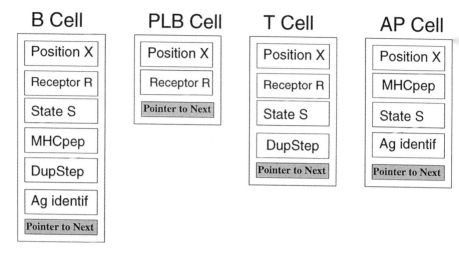

Figure 5: Cell attributes blocks

It is clear that we cannot afford such a static structure and that a more flexible, although more complicated, *dynamic memory* allocation technique is in order.

Having chosen the attributes for each entity (see Fig 5) the information required for each cell is organized in blocks of variables, *one block for each cell.* In this way, each cell gets a clear identity (record) in memory and can be easily told apart from the others.

These blocks are linked in a forward list, one for each class of cells (B List, T List, PLB List, APC List).

The lists are initialized at startup time and are dynamically managed at run time. When a cell dies out it is removed from the list. Similarly when a virgin cell comes in (from the bone marrow, lymph nodes and/or for clone division of stimulated cells) memory is allocated, the variables-attributes are filled with appropriate values and the whole block becomes the head of the list.

The memory grows linearly with the number of cells, following the clonal growth of cells during the immune response.

As a result, a look at the figures 2-3 reveals bursts in memory requirements right after the injection of antigens.

Fig 7 shows the amount of memory globally allocated for different initial number of cells introduced at starting time.

In Fig 7 it is showed the amount of memory required both initially and

T cell list

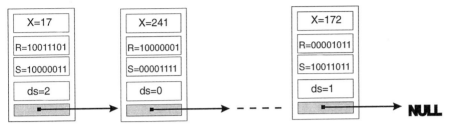

Figure 6: Example of list of cells

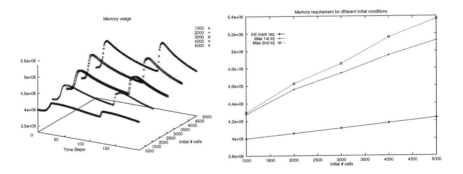

Figure 7: Dynamic memory requirement

after the first and the second response as a function of the initial number of cells.

By creating a list for each cellular entity, we prevented the combinatorial explosion of memory requirement for cellular data structures. The problem still remains for the molecules. It is not possible to assign a unique identity to each antibody because it would be useless to distinguish between two functionally identical molecules Moreover, the molecules (antibodies, antigens and immuno complexes) are much more numerous than cells.

In practice the molecules are represented as before, that is, lumped together.

The computational space complexity still remains exponential with the bit string length ℓ, at least as far as the antibodies and antigens are concerned.

3.2   State flags

The cellular entities need to be distinguished during the various stages of the response. The behaviour of a cell is determined by its *state*. For instance, a B cell can be found in any of the following states: *Active, Internalized, Exposing* and *Stimulated*. It might also have bound an Ab, in which case the processing is quite different from Ag processing. As a consequence, we need to keep track of this information as well. Finally B cells can be of *Memory*-type or not.

The following representation provides us with the possibility of adding new features by packing all the information in a *single byte* using a bit-flag format:

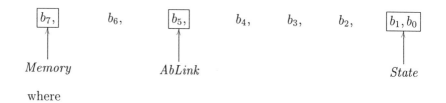

where

$$b_7 = 1 \Longleftrightarrow \text{is a memory cell}$$

$$b_5 = \begin{cases} 1, & \Longleftrightarrow \text{has bound an antibody molecule;} \\ 0, & \Longleftrightarrow \text{has bound an antigen.} \end{cases}$$

$$b_1 b_0 = \begin{cases} 00, & \Longleftrightarrow \text{Active state ;} \\ 01, & \Longleftrightarrow \text{Internalized state ;} \\ 10, & \Longleftrightarrow \text{Exposing state ;} \\ 11, & \Longleftrightarrow \text{Stimulated state .} \end{cases}$$

The remaining b_6, b_4, b_3, b_2 are left free for future use; 256 possibilities still leave much room for future developments.

This flag record is managed by a set of bit-mask operations, which allow to inquire the state of each cell (e.g. IS_MEMORY(), IS_ACTIVE(), AB_LINK()).

4   The parallel version of the simulator

The parallel version of ImmSim has been developed in two phases. In the first and preliminary version of the code (ParImm1.0) we adopted a simple master-worker scheme. This was good enough from the performance viewpoint but its storage requirements prevented from running large simulations, namely the very motivation behind the development of a parallel version. Nevertheless it is useful to report here some details about ParImm1.0 to allow a better understanding of the choices we made for ParImm2.0.

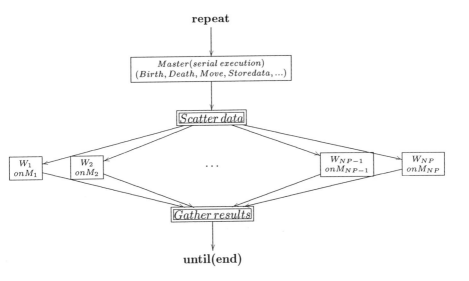

Figure 8: Master-Slave schema in ParImm 1.0

4.1 Parallel Immsim 1.0

In the serial ImmSim most of the time is spent in computing the interaction among entities belonging to the same site (*insite*). For ParImm1.0 we decided to run in parallel just that part, so there was a master and a set of workers. At each iteration:

1. the *master* distributed a subset of the sites $M_1, \ldots, M_{NP-1}$ to each *worker* $W_1, \ldots, W_{NP-1}$

2. *master* and *workers* carried out the simulation of the interaction among entities on their subset of sites

3. the *workers* sent back the results to the *master*

4. the *master* carried out all the other parts of the simulation (i.e. cells propagation, cells birth and death and so on).

During the execution of step 2 the tasks did not need to exchange any information. Actually the workers *never* talked each other. All the communication

302

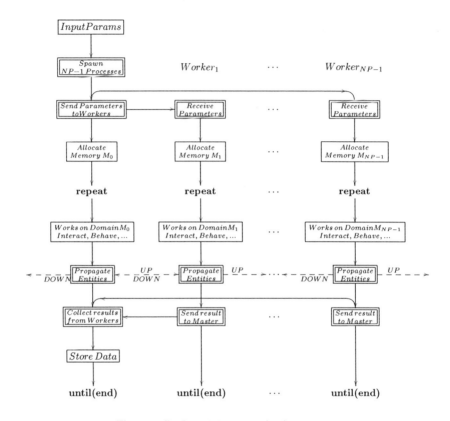

Figure 9: ParImm 2.0 communication pattern

was based on two basic collective operations:

- a *scatter* to distribute data (step 1)

- a *gather* to collect them at the end of step 3

Such communication pattern had the advantage of minimizing the changes with respect to the serial version of the code. Moreover the tasks executed just two, although pretty big (about 60 Kbytes per site), message passing operations per iteration. The communication phase was more bandwidth than latency bound and this was positive since our platform (the IBM SP2) is very suitable to such codes. However ParImm 1.0 shows all its serious limitations as soon as the issues described in 3.1 are considered. There we showed how the size of the huge data structure which represents the molecules grows exponentially with respect to the bit string length (ℓ).

In ParImm1.0 that static multidimensional array was not partitioned among the NP processors, as a consequence, simulations with longer bit string lengths did not show any *real* advantage running in parallel since the memory swapping activity of the master was and remained the actual bottleneck. So there was a compelling need to split the data structures among all the processors in order to avoid the memory swapping on the master node and allows, in such way, to run more realistic (i.e. with a longer bit string length) simulations.

We describe this solution in the following section.

4.2 Parallel Immsim 2.0

In the new version of the parallel code (ParImm2.0) *all* the phases of the simulation run in parallel. This major enhancement has been possible by using a particular technique for the *Domain Decomposition* of the cellular automaton space. The basic idea is to consider each subdomain as an independent entity which interacts with other independent entities. In other words there is no single list split among the processors but as many lists as the number of processors in use.

The body is mapped onto a bi-dimensional grid of $dim_X \times dim_Y$ sites, with periodic boundary conditions in both directions (up-down, left-right). However, to make the internal management of the lists easier, the sites of the automaton are not arranged as a two-dimensional array but as a linear chain. The transformation is carried out by a simple function $(X, Y) \to N$ and does not change the global toroidal topology of the body.

Each Processing Element (PE) is in charge of a number of sites

$$nsites = \frac{dim_X \times dim_Y}{NP}$$

When the division is not exact (i.e. there is a remainder) the extra sites are in charge of the first PE.

The assignment of the sites determines the ranking of the PEs with the exception of the first PE which always takes the first *slice* of the rectangular grid, (i.e. the sites from 0 to $nsites - 1$). In general the processor that works on the sites within the range $[k \cdot nsites, (k + 1) \cdot nsites - 1]$ is indicated as PE_k.

Each PE manages private lists of the B, PLB, T and AP cells belonging to its subdomain and the data structures describing its molecules as if they were the whole body. No PE keeps a copy of lists or data structures belonging to other PEs and in such a way for a fixed bit string length ℓ the memory required on a PE decreases linearly as NP grows.

Fig 9 shows the various subtasks of NP processes. The horizontal lines represent the communication among PE's.

Elements are deleted from or inserted in the lists of cells (and molecules arrays) when a subdomain boundary is crossed during the phase of diffusion. When a cell leaves the domain of a PE to migrate to one of the two nearest neighbour PEs, the corresponding element is deleted from the original list and all the relevant attributes are packed in a message sent to the PE which now owns the cell (PE_d). PE_d unpacks the message and inserts the attributes of the incoming cell in a new element that becomes the head of the list. Actually each PE packs all its outgoing cells in just two messages for performance reasons. The first is directed to $(PE - 1)$ mod NP, whereas the second goes to $(PE + 1)$ mod NP.

All the topology details and the dependence on the diffusion speed are confined in a single function which, for each cell and molecule of the body, takes in input the starting position and returns the new one. Such a choice facilitates the inclusion of two important features in the simulator: different diffusion speeds of distinct entities (e.g. since molecules are soluble they move much faster) and the support of different tissues (and/or organs) by means of simple changes in the topology of the body.

4.3 ParImm2.0 results

Both ParImm1.0 and ParImm2.0 have been realized, in practice, with the primitives for parallel programming defined by the PVM software package.[17] The main advantage of PVM is, for our purpose, the very simple mechanism for packing multiple and heterogeneous elements in a single message. Another important point is its availability on almost everything from laptops to vector supercomputers.

The results reported here have been produced on an IBM SP2 equipped with the High Performance Switch (HPS), a communication technology ables to deliver up to 40 Mbytes/sec. A proprietary version of PVM named PVMe[18] is available to take fully advantage of the HPS. Up to 16 *wide* nodes (each node being equivalent to a RISC/6000 model 590 workstation) have been employed for the tests.

All the timings we report are the total elapsed time of the simulation including all the disk I/O activity. Actually the initialization and termination of the parallel tasks are not included in the total time since they do not depend

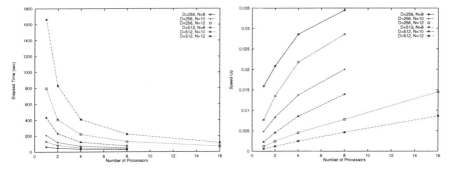

Figure 10: Time and SpeedUp $(1/T(NP))$ for several test cases

on the parameters of the simulation. For a fair comparison the single processor timings are obtained with the serial version of the simulator.

A number of runs with different grid size and bit string lengths (ℓ) have been executed. In figure 10 we present the results for two grids (16×16 and 16×32) and three values of ℓ (8, 10 and 12) on 1, 2 ,4, 8 and 16 PEs. To keep constant the concentration of cells in a single site the total initial number of entities is doubled going from the small (16×32) to the large (16×32) test case.

¿From figure 10 it is apparent that the execution time increases linearly with respect to the size of the automaton whereas a longer ℓ determines an exponential growth. We knew that the average message size (140 kbytes) was large enough to absorb the detrimental effects of the latency so we forecasted a good communication performance. Instrumented runs confirmed that the rate of the messages exceeds 30 Mbytes/sec.

Such communication rate along with a favourable ratio between computation and communication explains the pretty good speed-up observed in particular for the largest simulations ($\ell = 12$). Since in ParImm2.0 **all** the parts of the simulation run in parallel a good scalability should be achievable on any number of PE provided that ℓ is long enough.

A look at the results produced with the large grid (16×32) and $\ell = 12$ (figures 11) confirms the behaviour described in section 2.1. However the number of varieties (2^{12}) allows to gain a better insight into phenomena like the *hypermutation of antibodies.* [19]

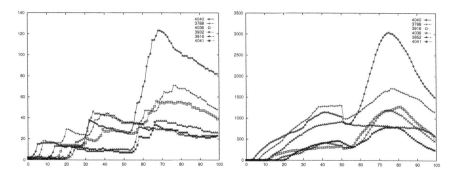

Figure 11: Big test case (16 × 32 and 4096 varieties)

Figure 11 is obtained from a 100 time steps simulation with 5000 initial cells population. The injections of antigens are scheduled for time step 0 and 50 and minmatch is settled to 10 over 12 bit string.

5 Conclusions and future work

This work sets the stage for several future investigations. In particular, we plan to look at the problem of affinity maturation of the antibodies and hyper-mutation of antibody genes in the simulation of the humoral immune response, two subjects of great interest to current immunological research. Two particular areas call for specific study. First, one would like to understand the role of mutations in the complement/determining region of the antibody, vis-a-vis structural mutations in the rest of the antibody. Are all regions of the antibody equally likely to mutate, or are there hot/spots instead? Are most mutations lethal or do they continue to exist in the system? What is the ratio of favorable to unfavorable mutations? The second area of interest is the germinal center. A lot of B-cell growth takes place in germinal centers, localized areas in the lymph nodes. Does all B-cell growth occur there? Is mutation and cell division synchronous or is there cycling of these phenomena as cells circulate in the germinal center? How are affinity cells selected?

Using the present parallel version of the Celada-Seiden model, we have the chance to test the various alternatives in order to predict what should be seen under different modes of cell stimulation and growth. The enhanced performance of the parallel code should allow our "in machina" results to be directly compared with "in vitro" and "in vivo" experiments.

Acknowledgements

F. Celada and P. Seiden are kindly acknowledged for many hints and valuable discussions.

References

1. P. Mezard, G. Parisi and M. Virasoro, *Spin Glass Theory and beyond* (World Scientific Singapore 1987).
2. F. Dyson, *Origins of Life* (Cambdrige Univ. Press, Cambridge 1985).
3. R. Mohler, C. Bruni and A. Gandolfi, Procs. IEEE **68** (Aug. 1980).
4. N. Vanderwalle and M. Auslos, in *Ann. Rev. of Comp. Phys. II*, D. Stauffer, editor (World Scientific, Singapore 1995) p. 45.
5. D. Stauffer and R. Pandey, Computers in Physicics **6**, 404 (1992).
6. M. Nowak, R. Anderson, A. Mc lean, T. Wolfs, J. Goudsmit and R. May, Science **254**, 963 (1991).
7. *Cellular Automata and modeling of complex physical systems*, P. Manneville *et al.*, editors, Springer Verlag Series in Physics n. 46, (Springer, Berlin 1989).
8. S. A. Kaufman, *The origins of order* (Oxford Univ. Press, New York 1993).
9. M. Kaufman, J. Urbain and R. Thomas, J. Theor. Biol. **114**, 527 (1985).
10. D. Stauffer and M. Sahimi, J. Theor. Biol. **166**, 289 (1994).
11. H. Meyer, Int. J. Mod. Phys. C **6**, 765 (1995).
12. R.J. de Boer, L. Segal and A. Perelson, J. Theor. Biol. **155**, 295 (1992).
13. P. Seiden and F. Celada J. Theor. Biol. **158**, 329-357 (1992).
14. F. Celada and P. Seiden, Immunology Today **13**, 56-62 (1992).
15. R. Pandey and D. Stauffer, J. Stat. Phys. **61**, 235 (1990).
16. O. Lefèvre, *IMMSIM* 2.1© *A User's Guide*, (Hospital for Joint Disease, NYU Medical School, New York, NY, 1995).
17. V.S. Sunderam, G.A. Geist, J. Dongarra and R. Manchek, Parallel Computing, **20** (1994).
18. M. Bernaschi and G. Richelli, Proc. of HPCN Europe 1995, B. Hertzberger and G. Serazzi, editors, (Springer, Berlin 1995).
19. F. Celada and P. E. Seiden, Eur. J. Immunology **26**, 1350-1358 (1996).

LIST OF PARTICIPANTS

Lecturers and Tutors

Franco Bagnoli, Dipartimento di Matematica Applicata, Università di Firenze, Firenze (Italy),

Ulrich Behn, Institute for Theoretical Physics, Leipzig (Germany),

Michele Bezzi, Dipartimento di Fisica, Università di Bologna, Bologna (Italy),

Franco Bignone, Istituto Nazionale sul Cancro, Genova (Italy),

Nino Boccara, Drecam/SPEC, C.E. Saclay, Paris (France) and University of Illinois in Chicago, Chicago (Illinois, USA),

Marcello Buiatti, Dipartimento di Genetica, Università di Firenze, Firenze (Italy),

Franco Celada, Università di Genova, Genova (Italy),

Giovanna Guasti, Arbeitsgruppe "Nichtlineare Dynamik", Universitæt Potsdam, Potsdam (Germany),

Raymond Kapral, Department of Chemistry, University of Toronto, Toronto (Canada),

Pietro Lió, Department of Genetics, University of Cambridge, Cambridge (UK),

Robero Livi, Dipartimento di Fisica, Università di Firenze, Firenze (Italy),

Jean-Pierre Nadal, Laboratoire de Physique Statistique, Ecole Nationale Superieure, Paris (France),

Michel Peyrard, Laboratoire de Physique, Ecole Nationale Superieure de Lyon, Lyon (France),

Arkady Pikovski, Arbeitsgruppe "Nichtlineare Dynamik" Universitæt Potsdam and Max-Plank Institute, Potsdam (Germany),

Antonio Politi, Istituto nazionale di Ottica, Firenze (Italy),

Stefano Ruffo, Dipartimento di Energetica, Università di Firenze, Firenze (Italy)

Philip E. Seiden, Thomas J. Wartson Center, IBM, Yorktown Heights, (New York, USA),

Dietrich Stauffer, Institute for Theoretical Physics, Cologne University, Koln (Germany),

Sauro Succi, C.N.R. - LAC, Roma (Italy).

Students

Livia Aromataro, Italy

Paola Babbi, Italy

Maria Barbi, Italy

Nokolaj Bernstein, Denmark

Ronald Bialozyt, Germany

Antonio Bonelli, Italy

Hans Martin Bröker, Germany

Luca Bucchini, Italy

Maria Serena Causo, Italy

Andrea Ciliberto, Italy

Simona Cocco, Italy

Jan Cupal, Germany

Roberta Donato, Italy

Robin Engelhardt, Denmark

Beatriz Gil-Maza, Spain

Peter Götz, Germany

Charlotte Korinna Hemelrijk, Switzerland

Karl Javorskj, Austria

Karen Lippert, Germany

Panagiotis Maniadis, Greece

Cateljine van Oss, The Netherlands

Janos Palinkas, Greece

Marina Piccioni, Italy

Marco Pierro, Italy

David Pugh, South Africa

Thophanis Raptis, Greece

Markus Rarbach, Germany

Alexander Renner, Austria

Dirk Tomandl, Germany

Maria Zakintinaki, Greece